KT-454-730

Changing Environments

edited by Dick Morris, Joanna Freeland, Steve Hinchliffe and Sandy Smith

wiley.com

in association with

The Open University

This publication forms part of an Open University course U216 *Environment: Change, Contest and Response*. The complete list of texts that make up this course can be found on the back cover. Details of this and other Open University courses can be obtained from the Course Information and Advice Centre, PO Box 724, The Open University, Milton Keynes MK7 6ZS, United Kingdom: tel. +44 (0) 1908 653231, e-mail general-enquiries@open.ac.uk

Alternatively, you may visit the Open University website at www.open.ac.uk where you can learn more about the wide range of courses and packs offered at all levels by The Open University.

To purchase a selection of Open University course materials visit the webshop at www.ouw.co.uk, or contact Open University Worldwide, Michael Young Building, Walton Hall, Milton Keynes MK7 6AA, United Kingdom for a brochure. tel. +44 (0) 1908 858785; fax +44 (0) 1908 858787; e-mail ouwenq@open.ac.uk

Copyright © 2003 The Open University

First published 2003 by John Wiley & Sons Ltd in association with The Open University

The Open University
Walton Hall
Milton Keynes
MK7 6AA
United Kingdom
www.open.ac.uk

John Wiley & Sons Ltd
The Atrium
Southern Gate
Chichester
PO19 8SQ

www.wileyeurope.com or www.wiley.com

Email (for orders and customer service enquiries): cs-books@wiley.co.uk

Other Wiley editorial offices: John Wiley & Sons Inc., 111 River Street, Hoboken, NJ 07030, USA; Jossey-Bass, 989 Market Street, San Francisco, CA 94103–1741, USA; Wiley-VCH Verlag GmbH, Boschstr. 12, D–69469 Weinheim, Germany; John Wiley & Sons Australia Ltd, 33 Park Road, Milton, Queensland 4064, Australia; John Wiley & Sons (Asia) Pte Ltd, 2 Clementi Loop #02–01, Jin Xing Distripark, Singapore 129809; John Wiley & Sons Canada Ltd, 22 Worcester Road, Etobicoke, Ontario, Canada M9W 1L1.

All rights reserved. Except for the quotation of short passages for criticism or review, no part of this publication may be reproduced, stored in a retrieval system or transmitted in any form or by any means, electronic, mechanical, photocopying, recording, scanning or otherwise, except under the terms of the Copyright, Designs and Patents Act 1988 or under the terms of a licence issued by the Copyright Licensing Agency Ltd, 90 Tottenham Court Road, London W1P 0LP, UK.

Library of Congress Cataloging-in-Publication Data

A catalogue record for this book is available from the Library of Congress.

British Library Cataloguing in Publication Data

A catalogue record for this book is available from the British Library.

ISBN 0 470 84999 1

Edited, designed and typeset by The Open University.

Printed by The Bath Press, Glasgow.

1.1

Contents

Series preface

Environment: Change, Contest and Response is a series of four books designed to introduce readers to many of the principal approaches and topics in contemporary environmental debate and study. The books form the central part of an Open University course, which shares its name with the series title. Each book is free-standing and can be used in a wide range of environmental science and geography courses and environmental studies in universities and colleges.

This series sets out ways of exploring environments; it provides the knowledge and skills that enable us to understand the variety and complexity of environmental issues and processes. The ideas and concepts presented in the series help to equip the reader to participate in debates and actions that are crucial to the well-being – indeed the survival – of environments.

The series takes as its common starting point the following. First, environments are socially and physically dynamic and are subject to competing definitions and interpretations. Second, environments change in ways that affect people, places, non-humans and habitats, but in ways that are likely to reflect differing degrees of vulnerability. Third, the unsettled, uncertain and uneven nature of environmental change poses significant challenges for political and scientific institutions.

The series is structured around the core themes of **changing** environments, **contested** environments and environmental **responses**. The first book in the series sets out these themes, and (as their titles indicate) each of the three remaining books takes one of the themes as its main concern.

In developing the series we have observed the rapidly changing world of environmental studies and policies. The series covers a range of topics from biodiversity to climate change, from wind farms to genetically modified organisms, from the critical role of mass media to the measurement of ecological footprints, and from plate tectonics to global markets. Each issue requires insights from a variety of disciplines in the natural sciences, social sciences, arts, technology and mathematics. The books contain chapters by a range of authors from different disciplines who share a commitment to finding common themes and approaches with which to enhance environmental learning. The chapters have been read and commented on by the multidisciplinary team. The result is a series of books that are unique in their degree of interdisciplinarity and complementarity.

It has not been possible to include all the topics that might find a place in a comprehensive coverage of environmental issues. We have chosen instead to use particular examples in order to develop a set of themes, concepts and questions that can be applied in a variety of contexts. It is our intention that as you read these books, individually or as a series, you will find thought-provoking and innovative approaches that will help you to make sense of the issues we cover, and many more besides. From the outset our aim has been to provide you with the equipment necessary for you to become a sceptical observer of, and participant in, environmental issues.

In the series we shall talk of both 'environment' and its plural, 'environments'. But, what do these terms mean? As a working definition we take 'environment' to indicate surroundings, including physical forms and features (land, water, sky) and living species and habitats.

'Environments' signifies spaces (areas, places, ecosystems) that vary in scale and are connected to, are a part of, a global environment. For instance, we may think in terms of different environments, such as the Scottish Highlands, the Sahara Desert or the Siberian taiga (forest). We may think in terms of components of environments, for instance housing, industry or infrastructures of the built environment. The phrase 'natural environments' may be used to emphasize the contribution of non-human organisms and biological, chemical and physical processes to the world's environment.

It is useful to keep in mind four defining characteristics of environment: environment implies surroundings; nature and human society are not separate but interactive; environment relates to both places and processes; and environments are constantly changing. These characteristics are merely a starting point; as we proceed we shall gradually reveal the relationships and interactions that shape environments.

There are several features of the books that are worthy of comment. While each book is self-contained, you will find references to earlier and later material in the series depicted in bold type. Some of the cross-references will also be highlighted in the margins of the text so that you may easily see their relevance to the topic on which you are currently engaged. The margins are also used in places to emphasize terms that are defined for the first time.

Another feature of the books is the interactivity of the writing. You will find questions and activities throughout the chapters. These are included to help you to think about the materials you are studying, to check your learning and understanding, and in some cases to apply what you have learned more widely to issues that arise outside the text. A final feature of the chapters is the summaries that appear at the end of each major section, to help you check that you have understood the main issues that are being discussed.

We wish to thank all the colleagues who have made this series and the Open University course possible. The complete list of names of those responsible for the course appear on an earlier page. Particular thanks go to: our external assessor, Professor Kerry Turner; our editors, Melanie Bayley, Alison Edwards and Lynne Slocombe; our tutor panel, Claire Appleby, Ian Coates and Arwyn Harris; and to the secretaries in the Geography Discipline, Michelle Marsh, Jan Smith, Neeru Thakrar and Susie Hooley, who have all helped in the preparation of the course. Last but not least, our thanks to our course manager, Varrie Scott, whose efficiency and unfailing good humour ensured that the whole project was brought together so successfully.

Andrew Blowers and Steve Hinchliffe
Co-chairs of The Open University Course Team

Introduction

Dick Morris

This book introduces some of the processes by which change occurs in different environments and considers the ways in which humans are involved in such changes. In doing so, it provides an introduction to some of the science, technology and social science that we can use to examine environmental change.

The first three chapters examine the nature of change, and look at some of the underlying biological and physical processes involved. Chapter One concentrates on the biological and geological processes that occur on the Earth largely independently of human activity, although humans may profoundly modify them. Chapter Two considers some of the ways in which humans interact with the non-human world, particularly by virtue of our access to fossil fuels and other energy sources that enable us to cause major changes in environments, of a different order from those caused by any other species. The importance of human interaction with the non-human environment means that we need to understand how and why human populations change, and Chapter Three explores the biology and history of, and some possible future trends in, human population numbers.

Using the basic knowledge and skills from these three chapters, the succeeding chapters examine the ways in which particular environments are changed by human activity and consider some of the consequences of these changes. Chapter Four looks at the way in which humans affect and are affected by the land, especially through our agricultural activities in providing food and other materials for human consumption. Chapter Five focuses on water, on the way in which we obtain supplies of fresh water and how we use the oceans to provide for our needs. Chapter Six concentrates on air and particularly on the way in which human activity is affecting the atmosphere and may be altering the climate. These first six chapters are largely concerned with biophysical processes, but human activity and environmental change are also strongly influenced by social, economic and political processes such as globalization, economic growth and political action. These form the topic of Chapter Seven. The Conclusion briefly summarizes the ideas from the preceding chapters, and draws out some general issues about the identification and measurement of environmental change.

Each chapter draws examples from different environments to illustrate environmental change over a wide range of scales of *time* and *space*. Time and space are essential concepts in our analysis, and they are used in conjunction with three key questions:

- How and why do environments change over time and across space?
- How do we determine the consequences and significance of changes?
- How can we represent, model and predict changes and their consequences?

These questions appear at intervals throughout the text, and provide a common theme for the different chapters, but they also raise some other issues that you should keep in mind. To be able to recognize change in environments, we need to be able to describe, and possibly measure, their features. Throughout the text, you will encounter descriptions of different environments, using words, pictures, graphs and numbers. In the widest sense of the word, these descriptions are *models* of different environments, not the environments themselves. Keep asking yourself how such descriptions were obtained, and how reliable they may be: while the authors will certainly not be deliberately trying to mislead you, there are always likely to be difficulties of measurement and different interpretations of environmental information. The Conclusion to the book draws on the chapters and on some new material to examine these questions.

This is one aspect of the concept of *uncertainty*, which is examined in greater detail in the third book in this series (**Bingham et al., 2003**).

Allied to the general issues of *how* to describe and measure environments, there is the issue of *what* makes up a particular environment (see also the Preface to the series). Colloquially, we often speak of 'the environment', as though it were one thing, but we also appreciate that there are many different environments. At a general level, we refer to the *built environment* or the *natural environment*, but we may also accept that your environment and my environment are likely to be different, embracing different physical, biological and social entities, areas and relationships. In this book we have chosen to look separately at the land, water and air as environments, although we recognize that they are all closely interconnected. What we have done is to choose a particular *system* in each case. Such a separation of the complexity of the world around us into different systems can be a powerful means of simplifying this complexity and making it apparently more understandable, but it can also have some unintended consequences, as we will consider in the Conclusion to this book.

Reference

Bingham, N., Blowers, A.T. and Belshaw, C.D. (eds) (2003) *Contested Environments*, Chichester, John Wiley & Sons/The Open University (Book 3 in this series).

Chapter One

Dynamic Earth: processes of change

Dick Morris and Charles Turner

Contents

1   Introduction

From an aeroplane, we can look down and see the landscape spread out below as a mosaic of different colours and shapes. Different countries and regions exhibit different but recognizable patterns. Much of England, for example, is characterized by a patchwork of fields, some arable, some under grass, separated by fences, hedges or walls. There are less regular patches of downland and rough grazing, usually on more sloping ground, whilst darker green or brown blocks indicate woods, heath or moorland. This is very different from, say, the pattern visible in the Alps but the elements of all these patterns correspond primarily to different forms, or assemblages of vegetation, that directly or indirectly support the diversity of other life: see Figure 1.1 (and Figure 1.15).

Had we been able to fly over many of these same areas even fifty years ago, they would have looked rather different. The further back in time it was possible to look at the scene, the more different it would look, and most of our surroundings are still changing and are never completely fixed. **Hinchliffe et al. (2003)** have stressed the importance to humanity of environmental change, so we need to understand how and why it occurs. This chapter looks at the biological and

(a)   (b)

(c)   (d)

Figure 1.1 Aerial photographs of different landscapes: (a) chalk downland (the Vale of the White Horse, Oxfordshire, England); (b) heather moorland (Ravegill Dodd, Scotland); (c) arable farmland (plain of the River Seine, Northern France); (d) mixed enclosed farmland and upland (Kirkhope Law, Scotland).

geological processes, involving interactions between living organisms and their surroundings, that have shaped our environments and will continue to do so into the future. It aims to show how processes involving living organisms, over very different scales of time and space, can cause change in environments. You will be able to use this understanding in later chapters to examine the ways in which environments change in response to human activities.

The chapter starts by considering the general nature of environmental change, and then examines the common processes that occur in all living environments. These processes are responsible for their continued function, but also cause them to change. The environments that we observe around us are also affected by processes that have occurred over geological time, and these form the topic of Section 5. The examples of these processes of change given in this chapter have been chosen to provide very contrasting, but complementary, understandings of our environments.

2   Environmental change in time and space

The word **dynamic** in the title of this chapter is used specifically to mean something that *changes with time*: different today from how it was at some point in the past, or is likely to be in the future. The opposite of dynamic is **static**: something that is static formally does not change, or at least appears not to change, over a certain time-period.

dynamic

static

To understand this idea of change better, let us consider some examples concerned with environments. An obvious example of something dynamic that springs to mind is things that move – water in a stream, birds flying past. Although I can't actually see it happening, the grass on my lawn keeps growing, leaves are beginning to sprout on the trees and I confidently expect those leaves to fall off in the autumn. The branches of the trees will elongate during the year, and over a number of years the trees will mature, possibly begin to die back and finally will fall, to rot down and disappear over a longer time, maybe to be replaced by others of the same or different species. The state of the village in which I live is also dynamic: people move in and out, new houses are built, shops close and traffic levels increase. A major concern at the beginning of the twenty-first century is climate change, probably associated with changes in the amount of carbon dioxide in the atmosphere. In human terms, the height and position of mountains and the shapes of continents seem static and unchanging. Yet we know that over periods of many human generations, all these do change and so are also dynamic.

I noticed several things about these examples. The changes involved a whole range of different properties and occur at different scales of time and space. The movement of water in a stream is a rapid physical change in spatial position of a given bit of water. The mass of the growing grass increases, as new tissue is formed by processes of photosynthesis and metabolism (which we will consider in more detail in Section 3).

Change can potentially be quantified in terms of numbers (of mass, species, distance and so on) but our reactions to change are also affected by our knowledge and values **(Belshaw and Hinchliffe, 2003)**. We may feel it is significant when a highly visible organism such as the tiger is threatened with extinction, but few of us are much concerned about the essential processes of birth, death and decay that change the numbers and mass of the organisms in the soil beneath our feet. To some extent, the attention that we give to change depends on the rate at which it occurs. **Hinchliffe et al. (2003)** give examples of environmental change that occur over the whole range of timescales from seconds to millions of years. We are generally much more aware of short-term, rapid changes than we are of changes that take several human generations or longer.

○　　From your general experience, give some examples of environmental changes that occur over time-periods of seconds, years, centuries and thousands of years.

●　　Changes over seconds might include the intensity of sunlight on a day with clouds and high winds or the position of a fast-moving animal or water flowing in a stream. Over years, you might have thought of the size of trees, the numbers of some animal species or the course of a river and the pattern of landscape. Over centuries, the shape of an eroding coastline changes, as do the form and way of life of human society as new technologies are developed. Change over thousands of years is much less obvious to an individual, and can only be detected through scientific measurements and their interpretation, but erosional changes in landform and major climatic change are moderately familiar examples.

Change also occurs as we move across space. As we travel around, we see different environments – such as the flat open fields, towns and villages, woodland – some of which are shown in Figure 1.1. These environments occupy different areas, such as a square metre, hectare (an area 100 m by 100 m) or square kilometre. In fact, our consideration of environmental change probably doesn't often concern areas as small as one square metre, although gardeners certainly manage environments at that scale.

○　　Identify environmental features and changes that you recognize and that concern you in an area 10 m by 10 m centred on where you live, in an area 100 m by 100 m and in an area 100 km by 100 km.

●　　From my perspective, some examples are:

　　　–　10 m: proximity of other dwellings; room temperature; family members; traffic; birds.

- 100 m: proximity of other dwellings; traffic; land use; people in general; shops and other facilities.

- 100 km: different forms of landscape, transport links. (I actually found it quite difficult to think of features at this scale. Yet I also worry about major global scale changes such as those in climate, perhaps at least as much as about local changes.)

To explain how and why environments change, we need to consider processes that occur at very different scales of time and space. In what follows, we are going to look at the very rapid, relatively small-scale processes of growth and decay of living organisms, at the longer-term interactions between different organisms and at the very long-term changes in physical aspects of environments that are almost imperceptible in human lifetimes.

Summary

- Features of environments that interest us exist on a wide range of different spatial scales.
- These features change with time, at a range of different rates.

Activity 1.1

Identify approximately the time and spatial scales involved in the following changes:

(a) growth of a leaf;

(b) growth of a plant;

(c) the change in temperature between summer and winter in northern Europe;

(d) the change in the landscape that is visible in travelling from London to Edinburgh;

(e) change in concentration of carbon dioxide in the atmosphere over the last 200 years.

(The answers to activities are given at the end of the chapter.)

3   Change in living communities

Figure 1.1 showed a range of recognizably different visible components such as heathland, different types of grassland or woodland, cereal fields. Each of these is characterized by a particular assemblage of forms and species, and corresponds to a distinct **community** of interacting organisms. The term 'community' is   community
sometimes used synonymously with habitat, defined by **Freeland (2003)** as the

'physical and biological environment in which an organism lives' but, for our purposes, community refers to the grouping of organisms that interact together to give a particular area its observed character. Although the overall pattern may be easy to see, defining the spatial boundary of a particular community can be difficult, indeed largely arbitrary, since they usually grade into one another. Individual communities occupy areas from a few square metres up to several square kilometres, and generally persist in recognizably similar form, sometimes for many years. How did these different communities arise, why do they persist, and what might cause them to change? In the UK one reason for the existence of different living communities is human activity. With agricultural land and forestry plantations, this is obvious, but even heather-clad moorland and 'ancient woodland' (which may have persisted without major clearance or replanting for at least 400 years) have been actively managed by farmers, gamekeepers, foresters and, now, by nature conservation organizations.

In fact, most of those communities that in the UK we regard as 'natural' require considerable human intervention to counter naturally occurring processes that would otherwise change them over time. In the 1920s and '30s the economic decline of British agriculture led to wholesale abandonment of many grassland and arable fields, which were rapidly colonized by hawthorn (*Crataegus monogyna*), blackthorn (*Prunus spinosa*) and other shrubs. A major clearance programme was required during the Second World War, when home production of food again became a priority: see Figure 1.2. In western Russia management of many of the state forests largely ceased during and for a while after that same war, with local

Figure 1.2 Chalk downland in England being sown during the Second World War. The scrubland in the background had developed following earlier changes in agricultural activity.

villages destroyed and much of the workforce killed or scattered. By the early 1950s, when management resumed, many tracks and paths had become completely impassable and dense scrub occupied previously open areas.

○   What common set of changes in the vegetation occur in both these examples?

●   Without human intervention, in both cases herbaceous (i.e. non-woody) plant communities were replaced by woody shrubs and ultimately trees.

A partly fortuitous experiment, which illustrates the way in which communities can change over a shorter period, and on a smaller scale, has occurred in the state of Bolivar, Venezuela (reported by Terborgh et al., 2001): see Figure 1.3. In 1986 the water level in Lago Guri was raised to create a larger reservoir for a hydroelectric power scheme. This isolated a number of previously continuous areas of forest as islands, varying in size from 0.25 to 150 ha. Investigation of the flora and fauna of these islands in 1993/4 showed that the smaller islands (<12 ha) had lost 75 per cent of the animal species known to occur on the adjacent mainland. In particular, all the larger animals that would normally prey on other animals were lost, and by 1996/7 there were wide variations in the numbers of other species between different islands: see Table 1.1 (overleaf).

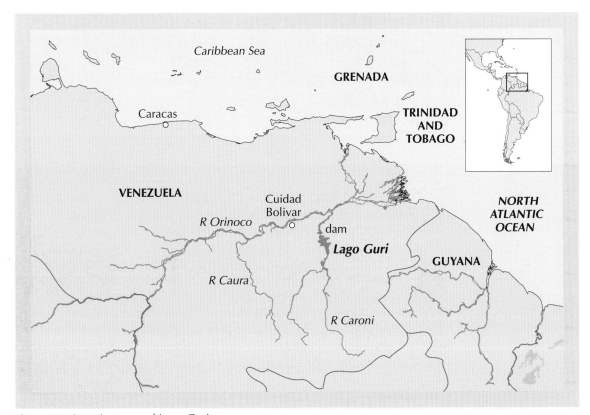

Figure 1.3 Location map of Lago Guri.

Table 1.1   Tree species diversity and densities of saplings, howler monkeys and mature leafcutter ant colonies on six small Lago Guri islands (1996/7)

| | Island | | | | | |
	Baya	Colón	Perímetro	Iguana	Cola	Palizada
Area (ha)	0.25	0.3	0.5	0.6	0.7	0.9
No. trees ≥ 10 cm DBH*	203	290	301	381	490	403
No. tree species	32	33	54	47	40	55
No. stems ≥ 1 m tall, < 1 cm DBH*/ha	780	2940	3120	1800	1680	6020
Howler monkeys/ha	4.0	10	0	10.0	8.6	0
No. leafcutter ant colonies/ha	4.0	6.7	4	6.7	4.3	1.1

Note: *DBH is diameter at breast height, about 1.4 m.
*Source:*Terborgh et al., 2001.

red howler monkey (*Alouatta seniculus*)

leafcutter ant (*Atta cephalotes*)

○   What major differences can you see among the data in Table 1.1?

●   Two of the islands had no howler monkeys, and the number of seedling trees (more than 1 m tall, but less than 1 cm DBH) varied between 780 and 6020 per hectare. The number of leafcutter ant colonies per hectare also varied from 1.1 to 6.7.

All these island areas were believed to be very similar prior to their inundation and separation, and no further human action occurred after this, but in a relatively short time the communities of organisms had changed greatly – and in different ways – between islands. Changes in any living community over timescales from hours to a few years are ultimately driven by what are called **trophic-dynamic** processes (broadly, what consumes how much of what) and **population dynamic** processes (how many organisms appear or die). We need to look at both aspects in order to understand how communities change, but we will concentrate initially on the trophic processes involving organisms in a food web **(Freeland, 2003)**.

trophic-dynamic

population dynamic

3.1 Organisms, energy and materials

The next chapter examines the concept of **energy** in more detail but, for present purposes, we need to recognize that all living systems require a supply of energy to survive. As a very crude analogy, living organisms are just like cars in needing fuel (energy) to operate.

○ How do we as humans obtain energy?

● Almost entirely from our food. The energy content of a foodstuff is often printed on the package, and is a major concern of dietary advice.

Animals need to eat bits of other living things to obtain energy, but the ultimate source of this energy is the Sun. Plants – from the smallest algal cell to the largest tree – take in solar energy and use it in the process of photosynthesis to form living material from simpler materials taken from their surroundings, as **Freeland (2003)** outlines. So, with the exception of some very specialized situations, plants are the basis of all life on Earth. The quantity of living material that they make each year is called **primary production**, and the plants themselves are **primary producers**. Tiny phytoplankton are the main primary producers in the oceans, wheat (and some weeds!) in a cornfield and trees in forests: see Figure 1.4.

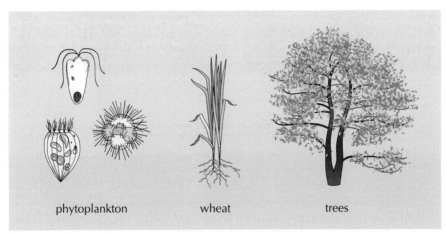

phytoplankton   wheat   trees

Figure 1.4 Some primary producers.

Because of their ability to 'feed themselves', plants (and a few other specialist species of particular environments) are classed as **autotrophs,** from the Greek for self-feeding. The rest of the living world comprises 'other feeders', or **heterotrophs. Freeland (2003)** introduced some different classes of these, categorized by what they consume: herbivores, such as cows, howler monkeys and ants, eat plants; carnivores, such as tigers, eat other animals. Perhaps less familiar categories are the **detritivores**, which eat dead plants and animals, and the **decomposers** (mainly bacteria and fungi such as the familiar field mushroom) which break down dead materials externally and absorb the

(margin notes) energy · primary production · primary producer · autotroph · heterotroph · detritivore · decomposer

breakdown products. In any food web there are different species of primary producers, herbivores, carnivores, detritivores and decomposers; these give the community its particular character.

To check that you understand these categories, pause for a moment and try to think of examples of each type of organism in familiar communities. When you have done so, look at Figure 1.5 which shows some different examples.

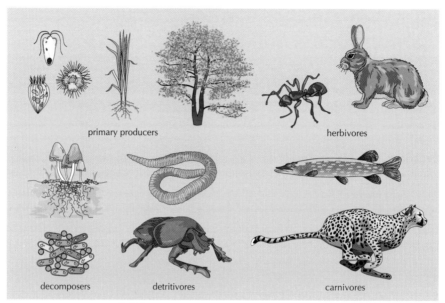

primary producers

herbivores

decomposers   detritivores   carnivores

Figure 1.5 Organisms fulfilling specific roles in different communities.

3.2 Ecosystem processes

Different communities can also be analysed as examples of ecosystems, defined as the interdependent groupings of plants, animals and other living organisms with non-living components, such as water, soil and minerals, that use energy and process materials (Tansley, 1935, quoted in **Blowers and Smith, 2003**). All ecosystems share a common structure and set of processes, where each of the different general trophic types can be represented in a simple flow-block diagram by a box, called an ecosystem compartment, as in Figure 1.6. The box labelled 'dead organic matter' represents the dead remains of living organisms which have not (yet) been consumed by detritivores or decomposers. This might be, for example, the dead heartwood standing in the trunks of living trees, the peat which accumulates in bogs or the **humus** and other dead organic matter in soil.

humus

The boxes in Figure 1.6 are linked together by arrows that represent flows of materials, energy or both between them. Starting with the primary producers such as plants, we can trace the movements around the system. The grass on a roadside verge for example, absorbs light and grows in size as it produces new leaves and other tissues. These tissues are built up from simple raw materials, including carbon dioxide from the air and water from the soil. A photosynthesiz-

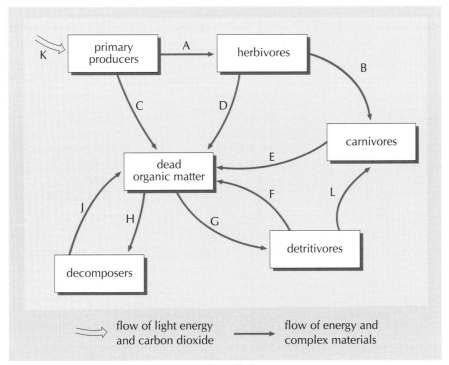

Figure 1.6 The basic compartments and interlinkages in an ecosystem.

ing plant in effect combines these materials together to produce carbohydrate (sugar is one familiar example, as is the starch found in cereal grains). This carbohydrate can then be **metabolized** within the plant in two ways. It can be combined with other raw materials from the soil to form the proteins, fats and so on, familiar from human dietary literature, that make up all living materials. At the same time, some of the original carbohydrate is used up in the process of **cellular respiration** (see Box 1.1) to supply the energy that is essential for all these other building processes.

metabolism

cellular respiration

Box 1.1 **Photosynthesis and respiration**

These two fundamental life processes are, in effect, mirror images. In photosynthesis, carbon dioxide is combined with water to produce a complex carbon-containing compound (carbohydrate) plus oxygen; in respiration, the complex compound breaks down, with oxygen being taken in and carbon dioxide and water released. The processes are essentially:

carbon dioxide + water $\longleftrightarrow$ complex carbon compounds + oxygen

In photosynthesis, light energy absorbed by plants drives the process from left to right. In respiration, the process goes from right to left releasing water, carbon dioxide gas and energy in different forms. (The different forms of energy will be examined in the next chapter.)

Building up new living materials requires energy; breaking them down releases it. Heterotrophs, such as ourselves, cannot photosynthesize, but rely on material from other organisms for both our raw materials and energy. Animal movement requires energy and energy is continually lost from organisms as heat and dissipated into the surroundings and out into space (as you will see in more detail in Chapter Six). Solar energy trapped through photosynthesis replaces energy lost to the ecosystem, so there is a continual *flow* of energy through any functioning ecosystem. In contrast, many of the materials involved are *cycled* around the compartments, as one organism feeds on another. On our roadside verge, rabbits and other heterotrophs eat some of the grass that has grown, but some of the leaves just die off, or perhaps get cut off and left there as part of highway maintenance. This dead material (along with any animals killed and left on the roadside) is then consumed by the detritivores and decomposers. This decomposer compartment is vital, since they complete the cycle by breaking down complex living materials into the simple forms that can be absorbed by autotrophs and built back up into living materials. Without them, all life processes would ultimately grind to a very messy halt!

○   Considered over a period of time, what limits the rate of flow of energy and materials between compartments?

●   Ultimately, the rates at which organisms take in materials, grow, reproduce and then either die or are eaten cannot exceed the rate of primary production, which effectively feeds the whole system.

zooplankton

The relative magnitudes of different flows can vary widely between systems. In marine ecosystems herbivorous **zooplankton** (tiny marine animals) may consume as much as 40 per cent of the primary production of phytoplankton, but on land most herbivores rarely consume more than about 10 per cent of primary production. The rest simply dies and ultimately decays. A typical deciduous tree like an oak may be habitat for a large number of insects and other herbivores, but such trees still have a nearly full canopy of leaves present at the end of the growing season, and these just fall and decay.

Activity 1.2

Identify which of the flows A–J in Figure 1.6 correspond to the following actions:

(a) a giraffe eating grass

(b) a dead leaf falling off a tree in autumn

(c) a mole eating a worm

(d) a worm eating a dead leaf.

(You will find the answers to activities at the end of the chapter.)

Figure 1.6 showed very generalized material flows within a stylized ecosystem, but there are important differences among materials in the details of the flows.

Some key links between the cycling of materials in ecosystems and change in communities depend on the cycles of the elements, carbon, nitrogen and potassium: see Box 1.2. Much of living matter is carbon, so local-scale carbon movements shown in Figure 1.7 correspond to the generalized flows in Figure 1.6.

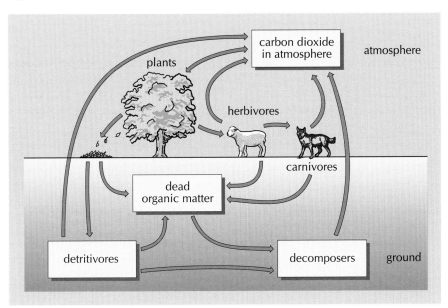

Figure 1.7 The local carbon cycle.

Box 1.2 Chemical elements

Every object on Earth is made of chemical **elements**. There are 109 named elements of which 92 occur naturally on Earth. The names of some are very familiar, such as carbon, oxygen, hydrogen, nitrogen, calcium, iron, gold, neon. For practical purposes, the smallest part of a chemical element that can exist as a separate entity is called an **atom**.

element

atom

Each element can be represented by a chemical symbol of one or two letters. This simply provides an abbreviated or shorthand form for writing the name. For example:

Element	*Symbol*
Carbon	C
Oxygen	O
Hydrogen	H
Nitrogen	N
Calcium	Ca
Iron	Fe
Gold	Au
Neon	Ne
Potassium	K

As you can see, some are simply the initial letter of the element's name. Where two or more elements have the same initial, a second letter is used for one of them – so, for example, C and Ca, N and Ne. (The first letter is always in upper case and any second letter in lower case.) In some cases the symbols are derived from Latin names which may seem rather obscure. So, Fe for iron comes from *ferrum*, the Latin word for iron; Au is from *aurum* meaning gold.

compound

Elements are usually found combined together to make **compounds**. There is an enormous range of possible combinations of different elements so there is an enormous number of different chemical compounds that

reaction

occur naturally or can be produced by chemical **reactions**, such as those in the metabolism of organisms. An individual particle of a compound is

molecule

called a **molecule** and in chemical reactions molecules may join together or gain/lose individual atoms to form new molecules.

The symbols of the elements are put together to give the formula of the compound. For example, hydrogen and oxygen combine to make water which has the formula H_2O. The number two indicates that 2 atoms of hydrogen combine with 1 atom of oxygen to make 1 molecule of water. (The subscript number only refers to the element that it follows, so for H_2O, the 2 only refers to hydrogen, not oxygen.)

Carbon dioxide is a compound of carbon and oxygen that has the chemical formula CO_2, meaning 2 oxygen atoms combined with 1 carbon. Carbon and oxygen can also form another compound, carbon monoxide. This has the formula CO and is made of 1 carbon atom and 1 oxygen atom.

(The reasons why 1 of this combines with 2 of that, or 3 of this with 1 of that and so on, are to do with the internal make-up of the atoms – you don't need to know any more for the purposes of this book.)

The chemical reactions in photosynthesis and respiration (Box 1.1) can be represented by the formula:

$$6CO_2 + 6H_2O \longrightarrow C_6H_{12}O_6 + 6O_2$$

You will meet this again in the next chapter.

○   In Figure 1.7, which flows of carbon represent photosynthesis and respiration?

●   The arrow from the atmosphere to the plant represents photosynthesis, and the arrows from plants, animals and decomposers back to the atmosphere represent respiration.

The so-called global greenhouse effect depends on the amount of carbon as carbon dioxide in the atmosphere, and this is affected by ecosystem processes. Figure 1.8 shows other processes involved in the global and longer-term

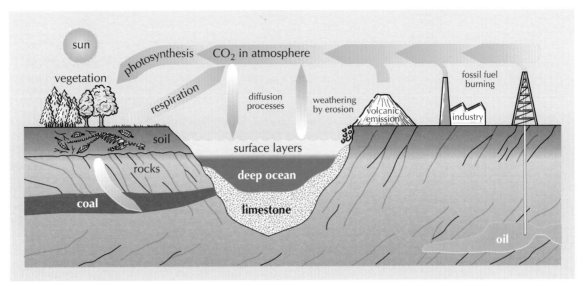

Figure 1.8 The global, long-term carbon cycle.

version of the carbon cycle. The local form is represented by the top left corner of this figure.

○   What are the major differences in spatial scale and timescale between the local and global aspects of the carbon cycle?

●   The local carbon cycle occurs within areas of at most a few square kilometres, over a matter of hours, although some of the carbon may remain in, for example, a tree for several hundred years. The global cycle involves movements across thousands of kilometres, over millions of years.

The next two figures, Figures 1.9 and 1.10, show two other cycles – for potassium, a component of the fluids in all living organisms, and for nitrogen, which is an essential component of proteins. Because biological, geological and chemical processes are involved, these are all termed **biogeochemical cycles** (as is the   biogeochemical cycle carbon cycle). Nitrogen, like carbon, can exist as a gas, whereas the potassium cycle only involves solid and dissolved states.

○   What distinctive feature of these two cycles can you see that has no direct counterpart in the short-term carbon cycle?

●   Both nitrogen and potassium occur in soluble forms in the soil, but the carbon in soil is shown as being present only as organic matter. In fact, there is a small amount of carbon dioxide dissolved in soil water, but its importance to the cycle is relatively minor.

Soluble material in soil can be taken up by plants or leached out by water flowing through it. If you are a gardener, you will appreciate that this is vitally important to ecosystem processes. Plants need nitrogen, potassium and some 20 other elements to grow. The rate of plant growth is always limited either by the amount

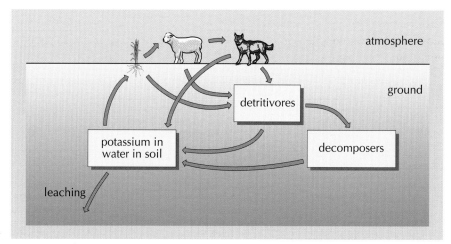

Figure 1.9 The potassium cycle.

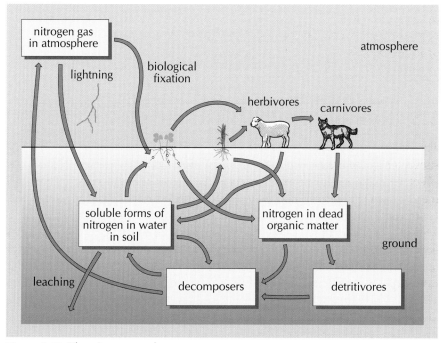

Figure 1.10 The nitrogen cycle.

of light energy that can be trapped by the leaves or by whichever of these necessary elements (nutrients) is in shortest supply. So, in a garden we encourage growth by adding fertilizers containing the necessary nutrients, or organic manure from which the nutrients are released by decomposition. In our absence, the rate of supply of nutrients from the soil depends primarily on the balance between decomposition of dead material and leaching by rain. Potassium that is leached into the oceans is effectively lost to land-based ecosystems for many millions of years, but leached nitrogen is mostly returned to the atmosphere after passage through a few more organisms and can return in a shorter time-period. Humans now interfere enormously in biogeochemical cycles through the production of fertilizers. The global rate of production of nitrogen fertilizer now equals or exceeds the rates at which nitrogen would otherwise move from the atmosphere into the rest of the cycle. This has major impacts on environmental change.

Activity 1.3

Fill in the gaps in the following paragraph with appropriate words from the list below. (Note: the answer to this activity provides a summary of this section.)

atmosphere, autotrophs, decomposers, detritivores, energy, organisms, solar

A biological community is made up of a range of .... For the community to function, there have to be .... present that can obtain .... from the Sun or other non-living sources. Most communities are dependent on .... energy trapped by plants, which may then be eaten by herbivores or parts may die and drop off to form litter. This is broken down by .... and .... The materials of which the living organisms are composed ultimately return to non-living forms, in the soil, in the .... or in the oceans.

3.3 Community interactions

The rate of primary production per unit area of land may be limited by nutrient supply, or by the area of leaf that is present to collect solar radiation. This can be affected by the number of herbivores grazing the leaves, and this may depend on the numbers of carnivores feeding on the herbivores. So there is a network of controlling interactions within any food web. So-called *bottom-up control* arises where nutrients or light limit the rate of primary production, and *top-down control* occurs through the effects of herbivores feeding on the primary producers and, in turn, of carnivores or detritivores feeding on them. The Lago Guri study (Terborgh et al., 2001), introduced earlier, gives an interesting insight into this.

○ On the islands left after inundation, small, young plants less than one metre tall were only about one-third as abundant as were plants of this age and size in the undisturbed forest. There, the density of herbivorous howler monkeys was approximately 0.3 per hectare, and there were less than 0.04 colonies of leafcutter ants per hectare. Do these data, and those in Table 1.1, suggest that the density of herbivores in the undisturbed forest was controlled (bottom-up) by the availability of food or (top-down) by the presence of predators?

● In most of the isolated islands, the densities of howler monkeys and leafcutter ants were much greater but there were many fewer young plants, compared to the undisturbed forest. This suggests that the higher numbers of monkeys and ants on the islands were

reducing the growth of the vegetation. This suggests that herbivore numbers in the undisturbed forest were controlled from the top-down by the presence of predators rather than bottom-up by the rate of growth of the vegetation.

Understanding what mechanism of control is operating in a particular situation may give guidance as to what changes are likely to occur. Ecosystem analysis concentrates on functional groupings of organisms and provides some general explanations of community function, but we are often more interested in the identity and numbers of different species. For individual organisms, ecosystem processes come down to the reality of obtaining energy and nutrients in order to grow and reproduce. There are limited amounts of materials or energy available in any ecosystem, so each organism may well **compete** with its parents, its siblings and with other species for those limited **resources**. In this context, the term resource refers to any material or energy source that is needed, and is used up by, any organism in order to survive. We will encounter a slightly different interpretation in relation to humans in later chapters. Each organism also has to compete with members of its own species for mates and, to avoid competition, organisms have mechanisms that enable them to **disperse** to new areas. Animals can move independently or be carried by wind, water or other animals, as can plant seeds. The couch grass (*Elymus repens*) that spreads its rhizomes through my garden is also dispersing to find light and nutrients. While obtaining resources for themselves, organisms need to avoid or minimize the chances of being eaten and thereby becoming the resource for another organism. Plants may do this by having defensive structures, as, for example, the thorns on roses. Some animals and plants produce antifeedant compounds that discourage others from eating them, and animals can fight or flee. The strong taste of mustard (*Brassica nigra*) and celery (*Apium graveolens*) arises in both cases from antifeedant compounds intended to deter caterpillars.

competition

resource

dispersal

couch grass (*Elymus repens*)

This picture of communities as being vicious, competitive places has often been invoked as a model for political and social processes, but we need to recognize that there are also many forms of interaction between organisms that are either mutually beneficial or at least neutral. One such interaction of enormous global significance is the symbiosis between plants such as clovers (*Trifolium* spp.) and nitrogen-fixing bacteria shown in Figure 1.10 and 1.11(a), whereby bacteria convert nitrogen gas from the atmosphere into a form that can be used by the host plant. In return, the bacteria receive carbohydrate and other materials synthesized by the plant. Some plants also avoid being grazed by providing a home for certain ant species that will attack potential grazers (see Figure 1.11b). So 'nature' is not entirely 'red in tooth and claw', as Tennyson claimed!

Activity 1.4

Summarize the different ways in which organisms within a community may interact. The answer to this activity forms a partial summary of Section 3.3.

(a)   (b)

Figure 1.11 Examples of positive interaction between species: (a) nodules of Rhizobium, a nitrogen-fixing bacterium, on French bean roots; (b) ant on thorn 'home' of Bulls Horn acacia tree, Belize.

3.4 Succession

Perhaps surprisingly, the results of all these different interactions between the multitude of different species in a real community show some consistent general patterns of change over time, referred to as **succession**. The changes that occurred after the inundation of Lago Guri are examples of what is called a **secondary succession**, from a pre-existing community structure, like the changes in the UK and Russia when agricultural or forestry activity was changed. An example of the alternative, **primary succession** from completely bare ground, occurred on the Krakatau island group of Indonesia. These lie in the Sunda Strait between Sumatra and Java, which are, respectively, 35 and 45 km distant: see Figure 1.12. In 1883 the main island, an ancient volcano, suffered one of the most violent volcanic eruptions of modern times. This reduced the island to less than half its original size and destroyed all the luxuriant vegetation and animal life that had previously inhabited it and the two smaller adjacent islands. It left behind a sterile unstable surface of lava and volcanic ejecta. However, recolonization of this new surface was rapid and by 1886 there was a luxuriant growth of ferns (presumably dispersed as tiny wind-borne spores from Sumatra and Java) with much of the ground surface covered by a thin layer of photosynthetic cyanobacteria.

succession

secondary succession

primary succession

○   Once seeds or spores reached the islands, what might limit their subsequent growth?

●   For plants to grow, they need water and nutrients in an appropriate form. The volcanic ejecta would contain negligible amounts of some of these, especially nitrogen. Unless some of the arrivals were able to fix nitrogen from the air, plant growth would be limited by the very low input of soluble forms of nitrogen in rain.

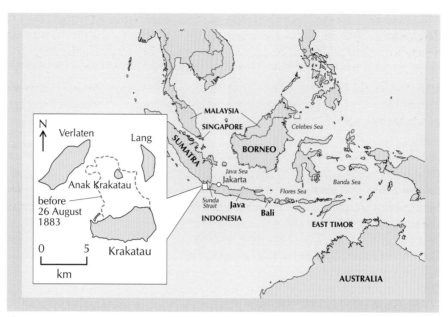

Figure 1.12 The location of Krakatau.

Table 1.2   The number of plant species recorded on the Krakatau islands at different times since 1886

Year	1886	1897	1908	1928	1934	1979*	Total known to have occurred
Total no. of plant species	26	64	115	214	271	196	387
No. of littoral plant species	10	30	60	68	70	75	–
No. of ferns	11	12	9	38	55	41	83

Note: * Data for 1979 relate only to the main island.

Table 1.2 shows the subsequent history of plant colonization of the Krakatau islands. In 1934 it was estimated that 41 per cent of the plants were probably wind dispersed, 28 per cent had floated in from the sea, including many of the littoral species (those of the beach or shoreline), a further 25 per cent had been carried in by animals, predominantly birds, and the rest perhaps by human activities.

○   Why should such a high proportion of the records be for littoral species?

●   Transport by currents is a major method of seed dispersal, but also beaches are a major feeding ground for birds, so seeds either defecated or carried attached to feet or feathers could easily arrive there. Finally, these are the areas most accessible to visiting botanists and so littoral plant species are likely to be well recorded!

Some species reached the island and were recorded but subsequently disappeared. This is an essential feature of succession. Initial colonization depends on some autotroph being able to reach the site and obtain the necessary nutrients from the solid surface, from rainfall or from the air in order to grow. At least initially, the supply of light is unlikely to be limiting. These **pioneer species** usually have easily dispersed seeds or spores and, ideally, can fix nitrogen. They generally have short lifespans, and germinate, grow, set seed and die within one season. (In other circumstances, we would recognize these as weeds!) The pioneers are usually not tolerant of shading by taller-growing species, so if any of these taller species arrive and grow they will tend to suppress or even eliminate the pioneers. The pioneer species and their successors form a food source for herbivores and, as all these organisms die, they return dead organic matter to the soil.

pioneer species

○   What will happen to this organic matter after the organisms die?

●   As you saw in the preceding section, decomposer organisms will break down the organic matter into simpler forms that can be used again by plants.

Some of the organic matter will be more difficult to break down and will remain in the soil as humus. The nature of the soil will be changed by this, as you will see in Chapter Four, and the soil will affect, and be affected by, the plants that are able to grow on it. The taller herbaceous vegetation will further change conditions above and below the soil surface, again influencing what can grow. Competition for light is often a major factor, so that perennial species that can carry their leaves even higher may have an advantage. But they also have the disadvantage that they need to produce a lot of non-photosynthetic (that is, non-productive) structures to carry their productive leaves up into the light. These structures need to be able to withstand wind and the depredations of animals, so are usually woody. So the next stage is indicated by the presence of scrubby vegetation – hawthorn and blackthorn in the UK, as we saw at the start of Section 3.

The final stage of this classic succession is the steady increase in size of the woody component, with trees growing up to form a more or less continuous woodland canopy. The whole process is indicated diagrammatically in Figure 1.13 (overleaf).

As succession proceeds, the total **biomass** (that is, the mass of living tissue) per unit area generally increases. The rate of primary production per unit area can increase in the early stages, as more and more of the available light energy falls onto leaves, rather than bare ground. Once the whole surface is covered with several layers of leaves, all the available light is being used, and primary production rate is at a maximum. As primary production builds up the biomass, all this living tissue has to respire to survive and there comes a point where the rates of photosynthesis and respiration are in balance. Total biomass cannot increase further although its species composition may continue to change.

biomass

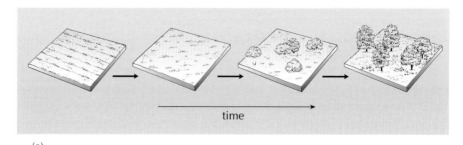

(a)

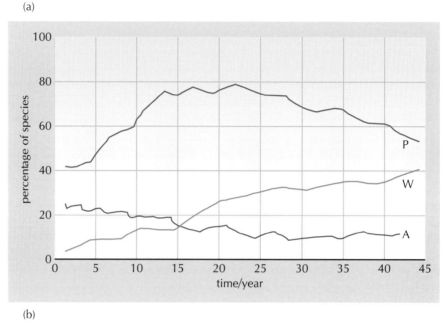

(b)

Figure 1.13   (a) The classic processes of succession from bare ground to forest; (b) changes during succession in the percentages of: A, annual; P, herbaceous perennial; and W, woody perennial species.

The pattern of nutrient cycling also changes with succession. In the early stages the system tends to be relatively 'leaky', and soluble substances are lost by leaching. As succession proceeds, more and more of the nutrient elements tend to be locked up in living or standing dead material and when any soluble material is released through the death and decomposition of an organism, there is another organism in place to take it up immediately.

At some point, many communities reach what appears to be a stable state, without easily discernible change over periods often of many years, although all the necessary trophic processes of photosynthesis, growth and decay must still occur. This apparent steady state is called the **climax community**, and is regarded as the 'natural' vegetation of the area. Climax communities such as tropical moist forests are often characterized by leak-free and rapid nutrient cycling, which is visible in the form of rapid plant growth. This led to the misapprehension that these forests could potentially be highly productive for agriculture. However, most of the nutrients in such areas are bound up in the

climax community

biomass, and so are easily lost if the forest is cut down or burnt. We will return to this, and to its consequences, in Chapter Four.

Within a climax forest, there is also a continual turnover of trees, as individuals reach the end of their lifespan and die, to fall and be replaced by other trees. These replacements may have been growing almost imperceptibly slowly in the shade of the mature trees, or they may be shade-intolerant but fast-growing species which are only able to germinate in the temporary patch of light caused by the treefall.

The number of species present in a community also generally increases throughout the succession process. Any species can only survive within a limited range of each of a whole set of different conditions. These might include, for example, temperature, moisture content of the air, the concentration of certain elements in the soil, presence of particular prey or pollinator species. This whole multitude of conditions is referred to as the **niche** of that species. A   niche species will only be found in a community if somewhere within it all the prevailing conditions fall within the appropriate limits for that species. Climax communities are usually structurally complex so provide a wide variety of potential niches. Temperate woodland is often stratified into canopy, shrub, herb and ground layer communities, each accompanied by its dependent fauna: see Figure 1.14.

Figure 1.14 The stratification of a temperate woodland community: Pendunculate or Common Oak wood (*Quercus robur*).

In tropical forest, stratification can be even more complex and one hectare of tropical moist forest in Panama may contain over 90 different species of large trees, and of Amazonian rainforest sometimes more than twice as many. Further north the most diverse temperate forests in the USA – the Great Smoky

Mountains of Tennessee – have only 30 species per hectare. Forests in Panama have 70 mammal species but beech–maple forest in the Great Smoky Mountains only 31 in total. In the far north there are fewer forest mammals still, with only 15 or 16 species in Alaska. There are various theories to account for these differences (14 separate hypotheses, according to Lawton et al., 1994!).

Each community is thus the result of processes that operate continuously over a range of timescales, from short (trophic) to long (successional). The outcomes of these processes will vary depending on, amongst other factors, the global location of the site and on small-scale, local variations in the underlying rock, the slope and aspect of the site and particular historic events. So, even in the absence of humans, the community on an area overlying chalk in Northern Europe will differ from that overlying hard, granitic rock a few tens of miles away (think of Dartmoor compared to the Wiltshire Downs) and from the saltmarsh community of the Blackwater estuary **(Blowers and Smith, 2003)**. The different conditions in these sites lead to different potential sets of niches and, hence, different communities.

Activity 1.5

The Ythan estuary in Aberdeenshire receives water from a largely agricultural river catchment. It is an important nesting and feeding area for a number of bird species and is a nature reserve. Until the mid-twentieth century the river channel through the estuary was regularly dredged to allow shipping through. In recent years bird numbers have declined, and the area of intertidal mudflat covered by photosynthetic algae has increased. On the basis of what you have learned so far, suggest some possible mechanisms that may account for these observed changes.

Summary

For a summary of the earlier subsections look back at the answers to Activities 1.3 and 1.4. The following points summarize sections 3.3 and 3.4:

- Change in communities involves trophic processes, but these are controlled and modified by other effects. These include:
    - feeding
    - competitive or positive interactions between organisms
    - dispersal.
- Each species can only survive within a limited set of conditions (its niche), and these conditions can change as a result of interactions between species to cause a process of succession.
- Succession shows some consistent features and may lead to an apparently stable climax community.

4   Global vegetation patterns

The various processes described in the preceding sections interact with variations in climate and other features across the globe to give some consistent general patterns in vegetation communities or biomes. **Freeland (2003)** discussed the importance of temperature and rainfall in determining these, and Figure 1.15 illustrates the climax vegetation that would be expected in a model continent, whose shape represents the typical shape of the actual continents on the Earth. These taper off to the north and south, but their bulk tends to be more concentrated in the Northern hemisphere.

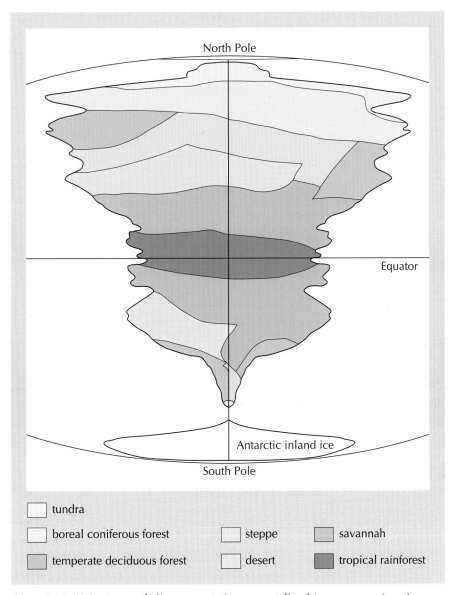

Figure 1.15 Major types of climax vegetation on a stylized 'average continent'.

bracken fern (*Pteridium aquilinum*)

common reed
(*Phragmites australis*)

cosmopolitan distribution

taxon

endemic

As you can see from Figure 1.15, at least in the Northern hemisphere, the more extensive vegetation types occur in bands that approximately follow lines of latitude. This emphasizes that the natural vegetation is strongly influenced by climate, especially temperature and rainfall. In the Southern hemisphere, the 'average' continent tapers in width, so its coastal climate and vegetation are more strongly affected by the ocean and by mountain ranges than in the Northern hemisphere.

From Figure 1.16 it should also be apparent that the classic progression to a climax forest is not universal, and in steppes, savannah and desert more or less sparse herbaceous vegetation is the climax. This reflects mainly shortage of water that allows only those specially adapted species to persist, or frequent fires which prevent the growth of taller woody species. Grazing of savannah areas also helps to prevent the formation of woody vegetation.

○   Looking at Figure 1.15, is climax forest the dominant vegetation type?

●   No. The largest area is occupied by herbaceous steppe and savannah.

Figure 1.15 is, of course, greatly generalized. Altitude affects local climate, particularly temperature, which generally falls with increasing altitude. Thus if rainfall remains roughly unchanged, in the tropics you would expect to pass from tropical forest at the foot of a mountain, through a temperate-like deciduous forest, to a boreal-like coniferous forest, before emerging into a treeless zone with the appearance of tundra towards the summit.

The preceding sections have considered some of the current, short- to medium-term mechanisms that account for change and variation in communities over time and across the Earth. Although these processes account for some of the general patterns, they do not provide anything like a complete explanation for the list of species that we find at any one point. Very few species have attained anything approaching a worldwide **cosmopolitan distribution**. Those that have include humans (*Homo sapiens sapiens*) and the rats, mice, lice and other organisms which accompany them. A few plants with broad climatic tolerances and extremely good dispersal mechanisms, such as the bracken fern (*Pteridium aquilinum*) and common reed (*Phragmites australis*), are widespread in both tropical and temperate regions.

Many **taxa** (groups of organisms) have only a single and often very restricted area of distribution, and these are termed **endemics**. Such distributions may reflect the last refuge of a formerly widespread species, such as that of the ginkgo or maidenhair tree *(Ginkgo biloba)*. Today this species occurs naturally in only a few localities in China, and even these may have been planted and protected by humans, but it was formerly widespread across the Northern hemisphere:

Figure 1.16 Different forms of climax vegetation: (a) tropical rainforest (Manaus, Brazil); (b) savannah (Masai Mara National Reserve, Kenya); (c) hot desert (Sonoran Desert, Arizona); (d) boreal forest (Athabasca River, Jasper, Canada); (e) tundra (Kazakhstan).

see Figure 1.17 (overleaf). At the other extreme, a single area of distribution may reflect a newly evolved species.

Other taxa possess **disjunct distributions**, that is, they occur in two or more widely separated areas and are called **vicariant** species. The tulip tree *(Liriodendron)* has two present-day species, *L. tulipifera* in North America and *L. chinense* in China. Fossils of *Liriodendron*, however, are found in sediments

disjunct distribution

vicariant

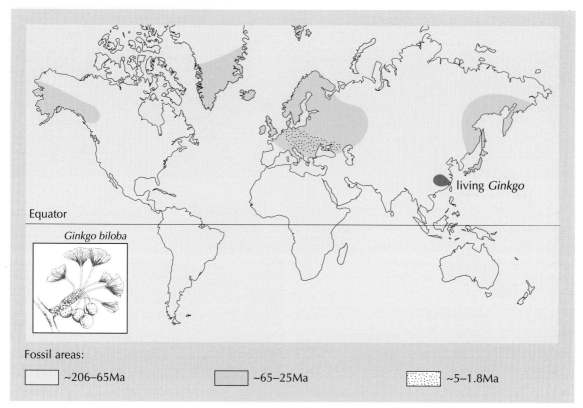

Figure 1.17 Current natural distribution of the maidenhair tree *Ginkgo biloba,* together with fossil occurrences of this genus.

Note: Ma = millions of years before present.

across Europe, clearly indicating that this genus was previously much more widespread: see Figure 1.18(a). The closely related genus *Magnolia* shows a similar distribution pattern with several species in eastern North America and others in China and also Japan: see Figure 1.18(b)).

Two different mechanisms have been invoked to account for disjunct distributions of both plants and animals. These are long-distance dispersal of organisms or some process that has reduced the area of occurrence of a species that once showed a continuous distribution over a wider area. The latter view is called *vicariance biogeography.* Both these mechanisms only come into play over much longer timescales than we have so far considered, so to explain and distinguish the causes of modern species distributions requires an understanding of geological and palaeoclimatic history, as well as of the evolutionary history of the taxa concerned. This brings out again the importance of time and space in considering environmental change, and takes us to the other extreme of change involving global scales and what are, to humans, unimaginably long timespans.

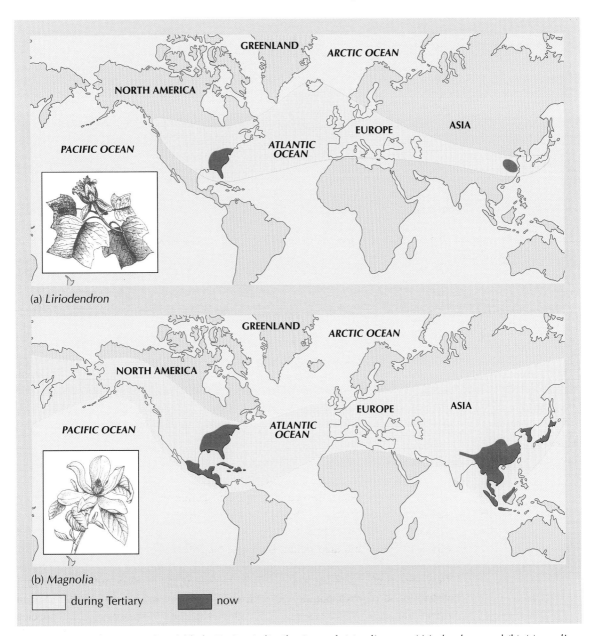

Figure 1.18 Current natural and likely Tertiary* distributions of: (a) tulip trees *Liriodendron*; and (b) *Magnolia* species.

Note: *The Tertiary period is shown in Figure 1.21.

Summary

- The processes of succession can lead to consistent patterns of vegetation (biomes), determined primarily by climate, and dependent on location on the Earth.

- Succession alone does not explain the precise mix of species that is found in any area, nor the distribution of species across the Earth. A few species are widely distributed, but some have very localized distributions.

- Where a given species occurs in very localized but widely separated areas, in order to explain this it may be necessary to look at processes occurring over much longer timescales than we have so far considered.

5   Solid but changing Earth

About 4,600 million years ago, during the formation of the solar system, gases and dust were gathered together under the influence of the Sun's gravitational field to form the planets including Earth. Eventually the semi-molten early Earth became differentiated into a solid inner core, a molten outer core, a solid layer called the mantle, and the solid crust, punctured by volcanoes: see Figure 1.19.

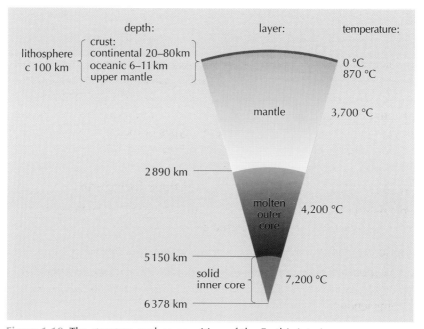

Figure 1.19 The structure and composition of the Earth's interior.

Gases brought to the surface by volcanoes gave the Earth its early atmosphere. It was certainly anoxic (almost entirely devoid of oxygen) but probably rich in carbon dioxide with low levels of nitrogen and water vapour. As the Earth cooled further, water vapour condensed into liquid droplets and was precipitated as rain to form oceans. The oldest rocks still preserved at the Earth's surface (in Western Australia, northern Canada and Greenland) have been dated at around

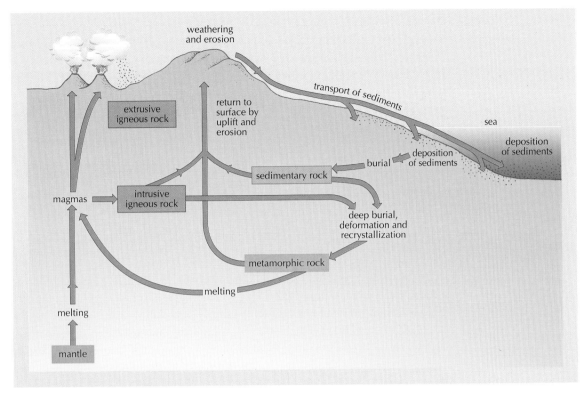

Figure 1.20 The rock cycle.

4,000 million years. However, they can be shown to consist of particles of older rocks, so that by that date the surface of the Earth had already undergone considerable change and had consisted of oceans and continental landmasses as it does today. The continuous geological processes of formation, destruction, transformation and reconstitution of different types of rocks – termed the **rock cycle** – were already happening: see Figure 1.20 and Box 1.3.

rock cycle

Box 1.3   Major groups of rocks and their formation

Geologists classify the rocks of the crust into three general groups, depending upon how they were formed:

- **igneous** rocks, such as granite, solidified from the molten state;

igneous

- **sedimentary** rocks, such as limestone, sandstone and clay, formed by the deposition of material from air, ice or water and its subsequent compaction;

sedimentary

- **metamorphic** rocks (of any origin) transformed (metamorphosed) by heat and/or pressure and recrystallized inside the Earth without melting: marble and slate are examples.

metamorphic

○   Of which of these groups of rocks must the crust have been made when it first formed?

●   Igneous rocks, formed by solidification of molten material.

The source of such molten material would be the mantle. Igneous rocks still form the bulk of the Earth's crust. All other rock types are formed from materials originally derived from igneous rocks. Look again at Figure 1.20 to trace these pathways. Over many thousands of years, wind and rain broke up the earliest igneous rocks into both soluble products and solid fragments – the processes of **weathering** and **erosion**.

weathering
erosion

○   What would you expect to happen to these fragments?

●   The smallest fragments would be carried by water, air or ice movement to be deposited as sediments in low-lying areas on land or in the oceans. There they could ultimately be compressed into rock by the weight of the overlying material. The sedimentary rocks most likely to be preserved were laid down in the oceans, or in river deltas, though sometimes lake deposits, wind-blown desert sand and glacial deposits have survived.

fossil

The sequences of rocks that have been formed over the last 4,000 million years provide us with a record of environmental change from the local to the global scale. This record is amplified, sparsely at first, but increasingly over the last 545 million years, by the records of **fossil** plants and animals. Fundamental to the understanding of this record is the *principle of uniformitarianism*, derived particularly from the work of the pioneer geologist James Hutton (1726–1797). The key observation made by Hutton was that the rocks of his native Scotland showed evidence of the results of essentially the same processes such as weathering, volcanic activity and changes in climate that are taking place on the Earth's surface at present. There are other processes, such as those taking place deep below the Earth's surface or in the ocean depths, which are beyond human sight, but where, ironically, it may be the evidence from the geological record that gives us vital clues to what is actually happening there today.

Geology became an established science in the nineteenth century, when geological mapping categorized knowledge of the spatial distribution of different kinds and sequences of rock units over the surface of the continents. This led to an understanding of their three-dimensional disposition and an ordering of rock units into sequences related to their time of formation.

○   Suggest two features of sedimentary rock sequences that might help identify their order of age.

●   (a) Where successive beds of rock are laid down as sediments, the uppermost will be the youngest (unless some subsequent major earth movements turn the whole sequence upside down!).
      (b) Many sedimentary rock units contain abundant fossils, including fossil plankton remains, pollen grains and spores, too small to be seen with the naked eye. Since there is a constant evolution of faunas and floras, experienced palaeontologists (scientists who study fossil remains) can use these to recognize the age of the rocks.

Originally it was only possible to determine the *relative* age of different rock units in the overall sequence, which was subdivided into the geological **eons, eras** and **periods**: these are listed in Figure 1.21. At least one may be familiar to you from popular culture! The actual age of geological periods in terms of millions of years was largely established in the twentieth century when various techniques of *radiometric dating* were developed that depend on the known, fixed rate of decay of radioactive isotopes in rocks.

eon
era
period

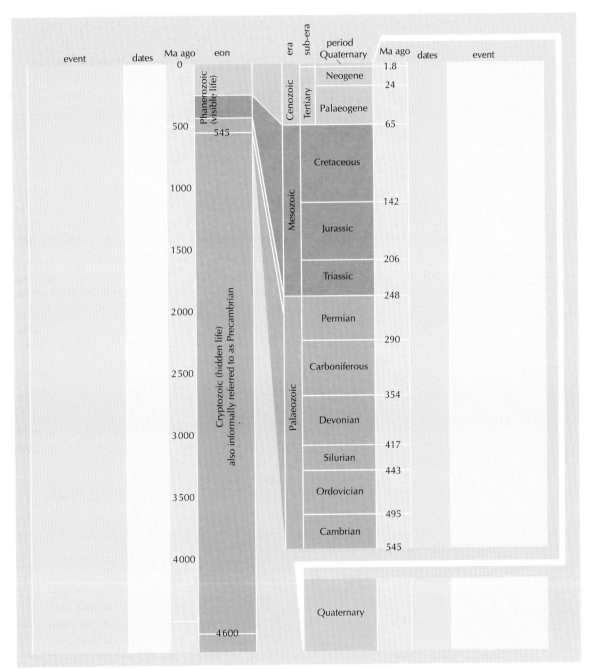

Figure 1.21   The geological timescale. (NB The blank columns are for use with Activity 1.6.)

Figure 1.21 should remind you that the history of the Earth is a very long one, measured in thousands of millions of years. To simplify the writing of these big numbers, we use the 'powers of ten' notation: see Box 1.4.

Box 1.4 Powers of ten

powers of ten

Writing out the noughts involved in large numbers such as thousands or, even worse, millions is tedious and prone to errors, so we usually express them as **powers of ten**. So 100 is 10×10, 1000 is $10 \times 10 \times 10$ and so on. We can write these in a standard form, so 10×10 is 10^2 (that is, ten-squared), $10 \times 10 \times 10$ is 10^3 (ten cubed, or 'ten to the power three'). When we get to numbers less than one, hundredths, millionths etc., which are the same as 1 divided by one hundred, one million or whatever, rather than writing them as $1/10^2$ or $1/10^6$, we replace the 1/ by a negative sign in front of the power. So, one hundredth is 10^{-2}, a millionth is 10^{-6} and so on.

- Using the powers of ten system, how would you write 1 million?

- One million is 1 multiplied by 10 six times over or '10 to the power six' and is written 1×10^6.

standard notation

That's fine for a single thousand, hundred or whatever, but we usually find ourselves dealing with numbers such as 1432, or 345.5. To write these in **standard notation**, we need first to locate the decimal point that appears after the last whole number and before the first part that is less than 1. That is the point between the fives in 345.5. There is an implied point after the 2 in the other example. We then 'move' the decimal point to the left, counting how many moves we need, until there is just one number to its left. So, three moves would make 1432 into 1.432. In effect, we have divided the original number by 1000, or 10^3 so we write the standard form as 1.432×10^3.

The commonly used multipliers actually have special names and symbols. Two common ones are m for milli- (division by 1000, a multiplier of 10^{-3}), as in the millimetre (one thousandth of a metre), and k for kilo (multiplication by 1000). Other multipliers are given in Table 1.3.

Table 1.3   Multiples of 10

Prefix	Meaning	Multiplies the fundamental unit by:	Power of ten	Symbol
micro-	millionth	1/1000 000 or 0.000 001	10^{-6}	μ[1]
milli-	thousandth	1/1000 or 0.001	10^{-3}	m
centi-	hundredth	1/100 or 0.01	10^{-2}	c
deci-	tenth	1/10 or 0.1	10^{-1}	d
kilo-	thousand	1000	10^3	k
mega-	million	1,000,000	10^6	M
giga-	thousand million	1,000,000,000	10^9	G
tera-	million million	1,000,000,000,000	10^{12}	T
peta-	thousand million million	1,000,000,000,000,000	10^{15}	P

Note: [1]This is the greek letter 'mu'.

Activity 1.6

Listed below is a series of important geological and biological events in the history of the Earth with their dates (sometimes approximate) in Million years (Ma). Enter the number for each event at its appropriate position in time in the columns provided on Figure 1.21 taking note of the geological period during which it occurred. Write in the dates given, using the powers of ten notation (e.g. 210 Ma is 2.1×10^8 years).

4,600 Ma		The formation of the Earth
3,500 Ma		Earliest evidence for life on Earth (bacteria)
550 Ma		Marine animals with hard skeletal parts appear
430 Ma		Major invasion of land habitats by plants and animals
c.220 Ma		Dinosaurs appear and diversify
210 Ma	2.1×10^8	Earliest mammals appear
100 Ma		Flowering plants (angiosperms) begin to diversify
65 Ma		A major meteorite impact results in severe environmental disruption and a mass extinction event.
2.6 Ma		Onset of the present Ice Age, which intensified about 0.65 Ma ago with intermittent glaciation of present-day temperate latitudes
c.2 Ma		Earliest known appearance of the genus *Homo*
c.0.2 Ma		Earliest known appearance of modern humans *Homo sapiens sapiens*
0.01 Ma		Beginnings of agriculture and farm animal domestication in the Near East
0.0005 Ma		European seafarers begin their global voyages

The answer to this activity is given in Figure 1.27 at the end of the chapter.

5.1 Moving the immovable: plate tectonics

Although we can recognize some contemporary changes in the solid surface of the Earth due to volcanoes, earthquakes and weathering, we tend to think of the general shape and position of the major continental landmasses as being relatively unchanging. The similarity of the general shape of the coastlines on either side of the Atlantic was noted by a number of early scientists and in the nineteenth century some 'catastrophist' geologists suggested that the Old and New Worlds had somehow been torn apart to occupy their present positions. In 1915 the German geophysicist and meteorologist Alfred Wegener first put together the scientific evidence for what he called *continental drift*, but his views were not widely accepted because a mechanism which could move the enormous mass of the continents across the surface of the Earth seemed incredible. However, in the mid-1960s a revolution in geological ideas occurred, and the new plate tectonics

theory of **plate tectonics** provided a framework within which many geological observations and processes, including those involved in global environmental and, indirectly, evolutionary changes, could be more easily understood. 'Tectonics' is the name given to the processes that involve the deformation and movement of solid rock within and at the surface of the Earth. Plates, in a geological sense, are explained below.

lithosphere
lithospheric plate

The outermost layer of the Earth, down to a depth of about 100 km or so and comprising both the crust and the uppermost part of the mantle (called the **lithosphere**: see Figure 1.19), is now understood to be divided geographically into a number of discrete parts called **lithospheric plates**, separated by zones of intense tectonic activity, where earthquakes occur. The plates are strong and rigid but rest upon a weaker and more easily deformable layer within the mantle. On a geological timescale, these plates can move around the surface of the Earth. The present distribution and boundaries of the major plates are shown in Figure 1.22.

mineral

The age of the crust varies over the surface of the Earth. In continental areas the oldest rocks were formed about 4,000 Ma ago, whereas in oceanic areas the crust is nowhere older than 200 Ma. The crust underlying the ocean basins, largely formed of igneous rocks, and rich in iron and magnesium silicate **minerals,** has a much higher density than the lighter continental crust. Most plates consist of areas of both oceanic crust and of continental crust fused together to form a single plate; others, such as the Pacific and Philippine plates, are mainly oceanic.

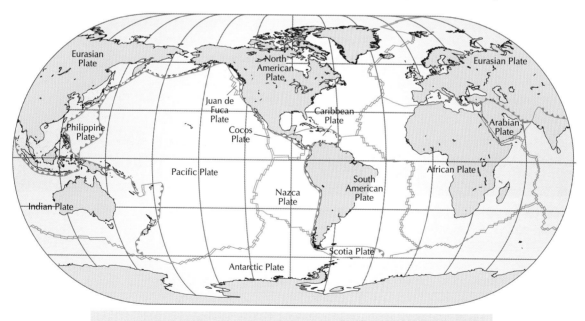

// divergent plate boundary ⌐ convergent plate boundary / transform fault plate boundary

Figure 1.22 The distribution of the major lithospheric plates, showing divergent, convergent and transform fault plate boundaries.

Oceanic crust is formed at **divergent plate boundaries** along **mid-ocean ridges**, from molten rock materials which rise from the mantle. The formation of new crust involves the two plates moving laterally away from the ridge at rates of a few centimetres a year in a process known as **sea-floor spreading**. As new ocean floor is created, any continental parts of the plates are carried away from divergent plate boundaries. At the moment the continents of South America and Africa, for example, are moving away from each other, widening the South Atlantic Ocean at the rate of a few centimetres per year.

divergent plate boundary

mid-ocean ridge

sea-floor spreading

Since the Earth is not increasing in size, the creation of new ocean floor has to be balanced by its destruction elsewhere at a **convergent plate boundary** where plates are in collision. Where two plates collide along a convergent plate boundary, if one has oceanic crust and the other continental crust, the lithosphere with denser oceanic crust is destroyed by being **subducted**, i.e. carried down beneath the other plate and back into the mantle. Sites of subduction are marked by the occurrence of deep **ocean trenches** where the water can reach depths of over 11 km: see Figure 1.23.

convergent plate boundary

subduction

ocean trench

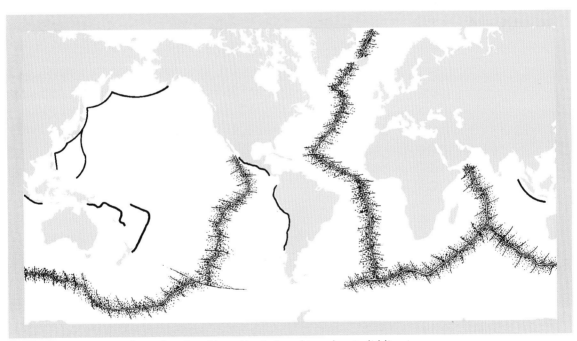

Figure 1.23 The ocean floor, showing ridges (shaded) and trenches (solid lines).

○   Convergent plate boundaries with subduction zones of this kind occur adjacent to much of the western coasts of both North and South America (see Figures 1.22 and 1.23). What obvious physical features characterize the land near these coasts? (Depending on your general geographical knowledge, you may need to use an atlas to answer this question.)

● There are chains of very high mountains, from the Andes in South America through Central America to the Rockies and, also in the western United States and Canada, the Cascades (which include Mount St Helens), which are the sites of many active volcanoes and very intense earthquake activity.

Where the denser oceanic lithosphere is subducted below the lighter continental lithosphere, the latter is thickened and a slow process of mountain-building occurs. The processes are too complex to discuss here but can also generate volcanoes and earthquakes. Plate movements not only affect the general landform of areas near plate boundaries, but also create uneven exposure to natural hazards such as earthquakes and volcanoes.

Where two plates with oceanic crust collide, one is subducted, forming a deep trench and chains of volcanic islands, or island arcs, such as the Caribbean Islands and the Aleutian Islands between North America and Siberia.

○ What might happen at convergent plate boundaries where two plates with less dense continental crust collide?

● They will not be dense enough to be subducted into the mantle, so they tend to crumple together and form mountain ranges. Collision of the African and European plates, for example, has resulted in the formation and uplift of the Atlas Mountains in North Africa.

transform fault plate boundary

When two plates move sideways past each other, this can lead to intense friction which, when released, results in earthquakes. This is known as a **transform fault plate boundary**.

Movement of the plates means that the current distribution of the continents is very different from what it has been in the remote past. Figure 1.24 shows some of these changes. Around 250 Ma ago, the continents were welded into a single supercontinent, Pangaea.

○ How has the latitudinal position of what is now the British Isles changed since the Devonian Period?

● In the mid-Devonian Period it lay in the Southern hemisphere, but by the Triassic Period it had moved north of the Equator and thence to its present latitude of about 55° North.

In the past, as today, the Earth had a tropical equatorial zone, flanked to north and south by desert belts. The geological record for north-west Europe suggests arid conditions in the Devonian and again during the Triassic (indicating it was in a desert-belt area), whilst warm tropical seas and coal swamps occurred around 300 Ma as the area passed through the equatorial zone.

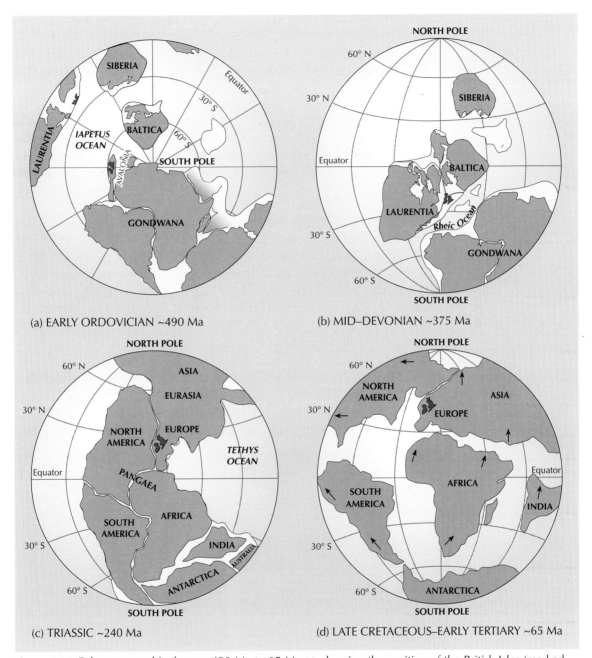

(a) EARLY ORDOVICIAN ~490 Ma

(b) MID–DEVONIAN ~375 Ma

(c) TRIASSIC ~240 Ma

(d) LATE CRETACEOUS–EARLY TERTIARY ~65 Ma

Figure 1.24 Palaeogeographical maps 490 Ma to 65 Ma BP, showing the position of the British Isles (marked in red).

5.2   Geology and life

These major changes in the position of particular continents on the Earth have not just affected the current form and nature of the rocks in a given area. They could and indeed must have had far-reaching effects on the living organisms present over time. Many of the currently existing taxa had already evolved when

Pangaea existed. Since then, oceans have opened up or been closed off; sea levels have risen and fallen; patterns of ocean currents and wind systems have changed as the position of continents and oceans altered; mountain ranges have risen and islands been created by volcanic activity or by separation from other landmasses.

Some of the anomalous present-day distribution patterns of related groups of plant and animal species discussed in Section 3.4 may be the result of changes that have taken place on these very different geological scales of time and space from the current ecosystem and successional processes we examined earlier.

Isolation of a major landmass over long periods can affect the organisms that are present there. Until about 100 Ma ago all the major southern continents were clustered together as the supercontinent Gondwana. As it broke up, one of the segments that drifted away was Australia (about 90 Ma ago), carrying with it a fauna including monotreme (egg-laying) mammals and marsupial (pouched) mammals but no placental mammals (i.e. those like cattle that suckle independent young): see Figure 1.25. These have evolved into the present very distinctive fauna of Australia, so different in its composition and evolutionary relationships from that of other continents. In fact, marsupial mammals were originally present in Cretaceous times on those other continents, including Europe, but except for a small number of species in South America and one very successful species, the Virginia opossum (*Didelphis virginiama*) in North America, they have been replaced on those continents by placental mammals.

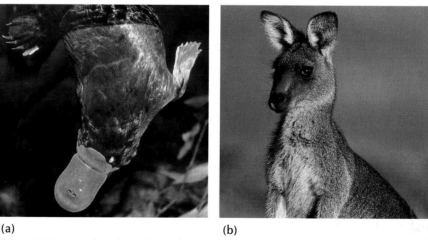

(a)   (b)

Figure 1.25 Examples of specialized Australian mammals: (a) a duck-billed platypus *(Ornithorrhyncus anatinus)*, a monotreme; (b) an eastern grey kangaroo *(Macropus giganteus)*, a marsupial.

Freeland (2003) shows that such highly specialized organisms are likely to be vulnerable to extinction in the face of species introduced from elsewhere in the world. This has indeed been the case in Australia, where introduced rabbits, cats and rats have contributed to major changes in the fauna and flora. Although small island populations are vulnerable to extinction **(Freeland, 2003)**, ancient islands may also allow **relict taxa** to persist which have now become extinct in the more competitive environments of the continents.

relict taxa

In the case of vicariant species of *Liriodendron* and *Magnolia* (which you looked at in Section 4), fossil evidence from pollen and leaf impressions suggests that their ancestors were present and widespread in temperate forests of Laurasia (the landmass of North America and Eurasia before it separated) during the Cretaceous. You can identify this in time and space from Figure 1.24. It is also known that those species were still present in western Europe (where they are no longer native) as recently as about 3 Ma ago in the late Neogene (Figure 1.21), but were probably no longer to be found in western and central Asia.

○ What two major plate tectonic 'events' from Figure 1.24 might have effectively created barriers isolating the populations of these plants in North America, Western Europe and Eastern Asia?

● An obvious barrier is the creation of the North Atlantic Ocean, as Laurasia was split up into North America and Europe/Asia by the creation of a divergent plate boundary, now represented by the Mid-Atlantic Ridge. Less obvious but equally important is the collision of the Indian plate and the Asian plate, which occurred after the stage shown in Figure 1.24(d), between 65 Ma and the present day, resulting in the great raised masses of the Himalayas and the Tibetan Plateau.

The Himalayas prevent rain-bearing wind systems reaching the interior of Central Asia, which is now an area chiefly of steppe from which the kinds of forest that would have supported *Liriodendron* and *Magnolia* have been eliminated. The repeated climatic oscillations of the Quaternary Ice Age **(Blowers and Smith, 2003)** in western Europe would have resulted in the regional extinction of these trees in western Europe.

Activity 1.7

The genus *Nothofagus* (the southern beeches) are forest trees of South America, New Zealand, parts of Australia and the islands of New Caledonia and New Guinea: see Figure 1.26. Clearly, this is a highly disjunct distribution pattern. There is also a good fossil record for the pollen of these trees for the Late Cretaceous (80 Ma ago). How would you interpret the present distribution pattern?

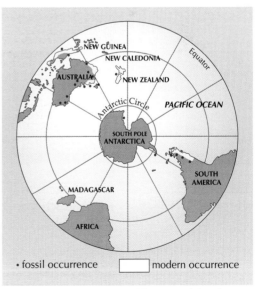

Figure 1.26 Distribution of living and fossil species of the southern beeches *Nothofagus*.

Summary

- The Earth has existed for around 4,600 Ma, and over that time there have been major changes in its atmosphere, surface features and the organisms that inhabit it.
- These changes, and the physical processes that have caused them, can be inferred from a detailed study of rocks and the fossilized remains of living organisms.
- It is now accepted that the processes of plate tectonics have resulted in global-scale movements of parts of the Earth's surface, before and since the emergence of living organisms.
- The opening up of oceans and the creation of mountain chains as a result of tectonic processes combine with the shorter-term evolutionary and ecosystem processes to determine the current distribution of organisms.

6   General summary and conclusions

We started this chapter considering the landscape of the UK and other areas of northern Europe as it is today. That appreciation is associated with spatial units within a range of sizes, and the appearance of any area strongly reflects the living communities of plants, animals and so forth that are present there. To understand how those communities are made up, how they function and how they came to be in the form they are currently, we have had to consider processes at a range of timescales, from the relatively short to many millions of years and over spatial scales from an individual leaf to the whole Earth. These include the current and continuously occurring ecosystem processes of photosynthesis, respiration and energy/material transfer, which keep communities alive, but can also result in their changing over time.

In addition to these basic ecosystem processes, which do not specifically consider the different species involved, there are other processes of interaction between different organisms that also affect the way the community functions. The result of all these processes is often a series of changes in communities over months or years that we identify as succession. The outcome of these successsional processes can be a community that appears to be in a steady state, although all the shorter-term processes and interactions still continue. The outcomes are greatly affected by the spatial location in which it takes place on the Earth, leading to sets of communities that typify particular areas.

However, these short- to medium-term processes, and the current spatial position of a community, do not fully explain the species that we find there. To explain these more fully, we need to consider change over very much longer periods and on a global scale. These processes are both geological, in the movement of the plates that make up the Earth's surface, and biological, in the evolution and extinction of species. Looking at timescales of millions of years helps to explain more aspects of current communities, but also reminds us that physical and biological changes have occurred in the past that are enormous compared to those that may occur during a few human lifetimes. However, as we shall see in succeeding chapters, the magnitude of human effects over a range of scales of time and space may be beginning to rival those of previous biological and geological effects.

You should now be aware of the major 'natural' processes of change in environments, and should be able to use this knowledge in evaluating the likely effects of human activity on environments.

There is another theme that has been implicit in this chapter, but has not been explored. This is the question of measurement of change. We have tended to assume that we can measure things unambiguously, but you should appreciate that time and space both greatly affect what and how we can measure. Identifying and counting the plants in one square metre can take a lot of time, as you will recognize if you have ever tried to do it. Counting animals raises other problems, including ethical ones concerned with what it is appropriate to do to them and simple logistical ones associated with their propensity to move around. To characterize numerically even a small environment is actually a major under-taking, and to repeat this sufficiently often to detect change is a costly exercise. We will be returning to some of these difficulties of measurement in Chapter Four and in the Conclusion, but you should now be sensitive to the fact that change in environments occurs over large areas and long timescales, so that the problems involved are not trivial.

References

Blowers, A.T. and Smith, S.G. (2003) 'Introducing environmental issues: the environment of an estuary' in Hinchliffe, S.J. et al. (eds) (2003).

Freeland, J.R. (2003) 'Are too many species going extinct? Environmental change in time and space' in Hinchliffe, S.J. et al. (eds) (2003).

Hinchliffe, S.J. and Belshaw, C.D. (2003) 'Who cares? Values, power and action in environmental contests' in Hinchliffe, S.J. et al. (eds) (2003).

Hinchliffe, S.J., Blowers, A.T. and Freeland, J.R. (eds) (2003) *Understanding Environmental Issues*, Chichester, John Wiley & Sons/The Open University (Book 1 in this series).

Lawton, J.H., Nee, S., Letcher, A.J. and Harvey, P.H. (1994) 'Animal distributions: patterns and processes' in Edwards, P.J., May, R.M. and Webb, N.R (eds) *Large-scale Ecology and Conservation Biology*, The 35th Symposium of the British Ecological Society with the Society for Conservation Biology, Oxford, Blackwell Science, pp.41–58.

Tansley, A.G. (1935) 'The use and abuse of vegetational concepts and terms', *Ecology*, vol.16, pp.284–307.

Terborgh, J., Lopez, L., Nuñez, P., Rao, M., Shahabuddin, G., Orihuela, G., Riveros, M., Ascanio, R., Adler, G.H., Lambert, T.D. and Balbas, L. (2001) 'Ecological meltdown in predator-free forest fragments', *Science*, 30 November, vol.294, no.5548, pp.1923–6.

Answers to Activities

Activity 1.1

(a) Leaf growth

Leaves grow and die over a period of weeks or months, and each occupies an area of something between a few square millimetres and a square metre.

(b) Plant growth

Whole plants can also grow and die over periods of weeks or months, but trees, for example, may take many years. Plants vary in size from, again, a few square millimetres to many square metres. Indeed, some single plants can occupy up to a hectare.

(c) Summer–winter

'Northern Europe' occupies an area of some 10 million square kilometres, and the basic changes of interest occur over periods of weeks and months. But this is complicated by the variation in temperature over much smaller scales of time and space.

(d) London–Edinburgh

The extent of change will depend very much on the scale at which we examine it. There are differences in the nature of the land surface and the organisms present over distances of a few metres and over timescales of weeks or months, but on a journey such as that suggested, we would probably notice only much broader scales of change. So, there are urban areas of a few square kilometres set in a larger area of 'countryside'. This countryside varies from the flat arable lands of the south east, through the more rolling mixture of grassland and arable in the Midlands, to the bleaker uplands of the Pennines and the Scottish borders. And all these areas change over the months and years, with the seasons and with development of human activities.

(e) Atmospheric carbon dioxide

This is a global phenomenon, but has occurred continuously, so, depending on how closely we can measure concentration, we may be concerned with change over only a few years rather than the whole 200-year span.

Activity 1.2

(a) Primary producer to herbivore (A).

(b) Primary producer to dead organic material (C).

(c) Herbivore to carnivore or detritivore to carnivore, depending on what the worm itself eats (B or L).

(d) Dead organic matter to detritivore (G).

Activity 1.3

A biological community is made up of a range of *organisms*. For the community to function, there have to be *autotrophs* present that can obtain *energy* from the Sun or other non-living sources. Most communities are dependent on *solar* energy trapped by plants, which may then be eaten by herbivores or parts may die and drop off to form litter. This is broken down by *detritivores* and *decomposers*. The materials of which the living organisms are composed ultimately return to non-living forms, in the soil, in the *atmosphere* or in the oceans.

Activity 1.4

Organisms interact in order to obtain resources, and these interactions may involve feeding on one another, or competing with others of their own or different species for limited resources of food, mates, territory etc., in order to survive and reproduce. However, some interactions, such as that between nitrogen-fixing bacteria and clovers, add to the resources available for both organisms.

Activity 1.5

There are several processes at work here. The silt brought down from the river would tend to be deposited in the estuary, changing its shape and probably changing the areas of mudflat that were present, especially since dredging had been discontinued. This might affect the area available for the birds to feed, but would not be expected directly to increase the growth of the algae. These might increase because they were not being grazed by so many birds, or they might be growing faster because of increased amounts of nutrients coming down the river (currently, the favoured explanation). Light income on the mudflats would rarely be a limitation on growth, unless the sediment being deposited was covering the photosynthetic tissue. The fact that photosynthesizing algae were increasing means that this is unlikely to be a factor.

Activity 1.6

The answer to this activity is shown in the completed version of Figure 1.21, reproduced here as Figure 1.27.

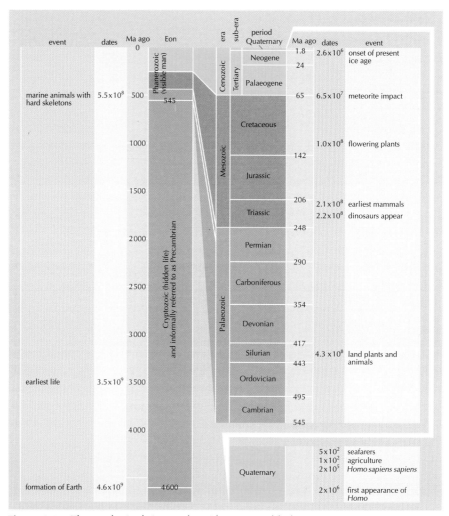

Figure 1.27 The geological timescale with events added.

Activity 1.7

The fossil record, which also shows that these trees once grew on Antarctica, before it was glaciated, suggests evidence for vicariance events. Australia, Antarctica and South America remained in close proximity until the beginning of the Tertiary. The fossil record demonstrates that *Nothofagus* was present in all three of these areas before that time, but it was absent from Africa which by 80 Ma ago was already far to the north.

Chapter Two

Dynamic Earth: human impacts

Alan Reddish

Contents

1   Introduction

Two quotations from the nineteenth century exemplify differing views of the relationship between humans and their environment. In 1859 Darwin wrote in *On the Origin of Species:*

> It is interesting to contemplate an entangled bank, clothed with plants of many kinds, with birds singing on the bushes, with various insects flitting about, and with worms crawling through the damp earth, and to reflect that these elaborately constructed forms, so different from each other, and dependent upon each other in so complex a manner, have all been produced by laws acting around us ...

In 1850 Tennyson, more vehemently, in *In Memoriam* had bitterly contrasted the cruelty of Nature with belief in a loving God:

> Man ...
> Who trusted God was love indeed
> And love Creation's final law –
> Tho' Nature, red in tooth and claw
> With ravine, shrieked against his creed.

(Note: 'ravine' here is an old version of 'rapine', i.e. cruel and violent exploitation.)

These are very different views, and you don't have to agree with either of them, however eminent their authors. But they do share a common human characteristic – the ability to express private thoughts, whether factual or emotional, in language that others may feel they can understand. Such communication plays an essential part in human co-operation of all kinds, often affecting the environment. Its importance is all around us, in our education, our workplaces, politics and the media. This chapter – indeed this whole series of books (see the back cover) – is part of this process, sharing in the belief that words can communicate one group's views about the environment to others. It is likewise open to your scrutiny and questioning.

This chapter continues the broad account of change on the Earth from the previous one, but now emphasizing human effects. It will show how the question:

- How and why do human activities change environments?

leads to the view that human use of additional sources of energy compared to other species has been a crucial factor, and hence to a second question:

- What is the nature of energy, and how does its use bring about these changes?

2   Humans on Earth
2.1 The last million years

The previous chapter concentrated on the changes on Earth occurring irrespective of human intervention, and contrasted two widely different timescales. On the one hand, there is the scale that fits in with our everyday experience of ecological processes – perhaps up to a century or so. On the other is the barely imaginable sweep of geological and evolutionary change over the thousands of millions of years we now believe to be the age of the Earth. Between these two is the subject of this chapter, an intermediate scale of a million years or so since the first appearance of humans. Even this is so far outside our experience as to require a considerable effort of imagination.

Homo erectus

Mammals, including the primates (apes and humans), have evolved over more than 100 million years, with humans differentiating from other primates over the last 10 million years or so. By a million years ago, there was a human-like species, *Homo erectus*, living in Africa. As outlined in **Blowers and Smith (2003)**, the whole time since then forms the present Ice Age, with glaciers covering large

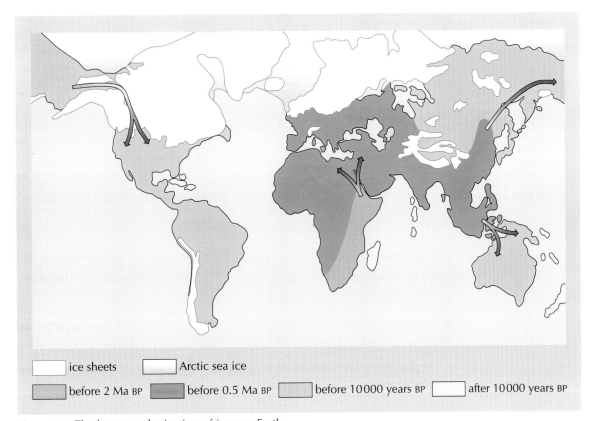

ice sheets   Arctic sea ice

before 2 Ma BP   before 0.5 Ma BP   before 10 000 years BP   after 10 000 years BP

Figure 2.1 The human colonization of ice-age Earth.
Source: Roberts, 1998, Figure 3.7, p.79.

areas of northern Europe and North America for part of the time, and slightly shorter Interglacial Stages, warmer periods more like the present, recurring every 100,000 years or so (see Chapter Six for more details). During this time these **'proto-humans'** had spread into Europe and Asia, with further evolutionary changes. By 100,000 years ago, modern humans – the species *Homo sapiens sapiens* – had evolved and had also spread through Africa, Europe and Asia. Most of the rest of the world was occupied by humans *before* the end of the most recent glaciation, about 10,000 years ago: see Figure 2.1. The *current* Interglacial Stage, the so-called Holocene, covers all the subsequent developments in human society that we can read about as ancient and modern history. See Box 2.1 for a way of representing these large timescales.

Box 2.1 Linear and logarithmic scales

This process of the history of the Earth can be summarized on a single **logarithmic scale**, in which equal intervals represent multiplication or division by ten, so the scale divisions are successive (in this case, decreasing) powers of 10, as explained in Chapter 1:

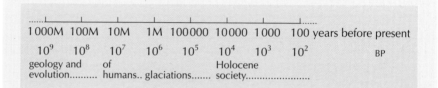

Logarithmic scales provide a useful shorthand way of covering very wide ranges of all kinds of quantities on a single diagram, but they can be deceptive if you are not used to them. We more often see quantities shown on a **linear scale**, which is much easier to interpret as in this case each interval on the scale represents the same amount. Staying with the example of time, it might be 100 years, here drawn with 1 cm for each interval:

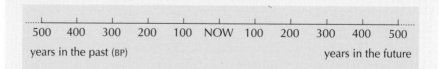

You can fill in dates from history on the left, or speculations about the future on the right, without difficulty about intermediate values: 150 is halfway between 100 and 200, 375 three-quarters way between 300 and 400 and so on. The problem is to represent very remote times on the same scale.

○   How far to the left would be 1000, 1 million and 1000 million years ago? (1 million years is abbreviated as 1 Ma.)

●   1000 years   5 cm to the left of the scale as drawn (10 cm from the centre)

1 Ma   1000 × 10 cm = 100 m to the left – well outside your room!

1000 Ma   1000 × 100 m = 100 km to the left – well outside your town!

So you can now see the advantage of the logarithmic version, in which all of these can be accommodated on the same page. The drawbacks are:

- Equal intervals on the scale represent extremely different time intervals.
- You never strictly reach 'now': the future would need a separate, increasing, scale.
- You cannot represent zero on a log scale, as positive and negative quantities need separate scales.
- The unevenness of the scale applies within the intervals as well, so it is more difficult to fill in intermediate values – this has to be done mathematically, or by using special graph paper. A rule of thumb is that 3 is shown approximately halfway between 1 and 10 and so on:

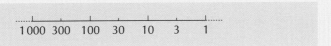

$$1\,000 \quad 300 \quad 100 \quad 30 \quad 10 \quad 3 \quad 1$$

We shall meet further examples of logarithmic scales later.

Activity 2.1

Before reading further, pause and ask yourself how human history and behaviour resemble, and differ from, that of other species. My suggestions follow in the next section.

2.2 Three strands in human behaviour

We can begin by distinguishing three strands, corresponding to vastly different parts of this history. The first is that large part of our inheritance **shared** with other living organisms, evolved over hundreds of millions of years. The second is the *peculiarly human*, whatever we may find it to be, that has differentiated our species from others for nearly a million years. The third is the *cultural* which has taken very diverse forms in the human societies that have developed over the ten

shared attributes

thousand years since the last glaciation. Our ability to communicate information, now over long distances, and to record it in books, films and so forth, means that we can pass it on from one region or generation to another without face-to-face contact being necessary. This has made possible a much more rapid process of lifestyle change than that associated with genetic evolution. During the Holocene our genetic make-up has hardly changed, but there has been a process of what is sometimes called **'cultural evolution'** with striking effects on the Earth. Changing methods of food production, social organization and technology have confronted famine and other natural disasters (catastrophic though these have often been for those involved), to support a still rising human population, now over 6,000 M. We should not lose sight of the total effect of this exceptional population growth on the other – non-human – inhabitants of the planet.

'cultural evolution'

M = million

Now let us look at these three strands in turn.

Shared attributes

It may be a matter of debate about *how much* of human appearance and behaviour is 'written in' to our genetic endowment, as represented by the DNA in all our cells, but it is at least an important part of the recipe. The universal role of DNA in distinguishing different species and individuals was introduced in **Freeland (2003)**. For the human case, the 'human genome project' set up to decode human DNA was able by 2001 to begin spelling out some of its incredibly complex detail (*Nature*, 2001).

The results have underlined the remarkable similarity between us and the rest of life. About 40 per cent of human DNA is similar to the DNA of such apparently diverse organisms as yeasts, simple plants, fruit-flies and worms – indicating the common basic 'machinery' of cell function and replication in organisms throughout the living world. Of course, there is an astonishing diversity both of form and behaviour in living organisms; its link with genetics may gradually become clearer. We may hesitate to identify human behaviour too closely with *particular* aspects of that of other species – survival and reproduction, mating rituals, social and hierarchical structures, co-operation, competition, signalling to intimidate intruders, defending territory, predation and so on. But, as with other species, our own genetic inheritance influences the way in which our behaviour emerges under our particular cultural and environmental influences. A good account of how earlier studies of animal behaviour ('ethology') can shed light on human philosophical questions is provided by *Beast and Man: The Roots of Human Nature* (Midgley, 1979), while more recently *The Perception of the Environment* (Ingold, 2000) raises questions about the way in which genetic and cultural influences are often separated in such discussions.

Distinctive human skills

It is not yet possible to identify the significance of the few per cent difference of our DNA from that of related species such as chimpanzees, as against the less than 1 per cent variation between different humans, or the extent to which these DNA differences relate to behavioural differences, but some general ideas seem clear. Certain human skills have evolved from those of other species and been developed, and the skills themselves are interlinked and mutually reinforcing; these have contributed to the exceptional environmental changes produced by human activities. I would like to suggest that four types of skill can be distinguished.

The first is manual (and bodily) *dexterity*, which enables us to interact with our environment with delicacy and precision. Some other animals have more agility, balance or speed, but typically human body–mind co-ordination enables humans to perform actions that are highly controlled and at a level of sophistication beyond that of other species (see Figure 2.2).

(a)   (b)

Figure 2.2 Dexterity in animals and humans: (a) weaver bird making nest (Cape Weaver bird [*Ploceus capensis*]); (b) human weaving basket (Gurung man, Nepal).

Activity 2.2

Can you think of other examples of human dexterity (excluding those involving the use of tools)?

(You will find the answers to activities at the end of the chapter.)

This dexterity is much enhanced by the second skill, the use of *tools*: again, some evidence of this is present in other species but what is distinctive about humans is the extent to which we have used some items from our environment to make tools (such as flint hand-axes, bows and arrows and ploughshares) to improve our ability to modify other parts.

Two particular cases are the use of rubbing sticks or flints to produce fire in a controlled way, and the use of projectiles to produce effects – usually damaging – at a distance. Both of these are examples of the distinctive use of *energy*, which I shall continue to emphasize as a defining human characteristic throughout this chapter, particularly in Sections 3 and 4.

Activity 2.3

What other examples of tool use, from hand-held to more elaborate ones, can you think of?

(You will find the answers to activities at the end of the chapter.)

From my answer, you will have appreciated that the gradual extension from simple hand tools to complex machinery has been a dominant feature of human society, and its ability to change the world.

Taken together, these first two skills of manual dexterity and tool manufacture/ use make up what may be called the 'craft' interaction of an individual with the environment, without the need for co-operation with anyone else.

The third skill, perhaps associated with the mental developments needed for tool use, and certainly more accessible to study, is the ability to *communicate* our thoughts about the world to others. Again, there is continuity with much animal behaviour – courtship rituals, alarm and threat calls, birdsong, whale, dolphin, chimpanzee communication. But there is also the gradual evolution of *conceptual thinking* (the use of abstract ideas going beyond instinctive responses) with a vast repertoire of communication means, from gesture, image-making and music to the central ones of words and of tokens such as money. All of these provide **models** to represent various aspects of the world – not only bare description, but also subtle logical, emotional or ethical responses to it. (Remember Darwin and Tennyson above.) It should perhaps be stressed that using the word 'model' for *all* these forms of communication goes beyond the everyday usage of this term for small-scale copies, or fashion icons, and is not universal. Some prefer to limit this technical sense to the often mathematical descriptions of complicated processes – such as the climate – which can be investigated using computers. But I prefer this broader usage to emphasize that *everything* we communicate about the world is some sort of representation, and is not the world itself. These models become so natural to us, as physical tools do, that we easily confuse them with the reality they represent.

model

The extent to which these communication skills have dominated human history can hardly be overstated, making possible all kinds of co-operative action to change the world. The further stage of *recording* words, images and sounds – in writing, books, paintings, discs, films and now the whole range of communication 'media' – vastly extends face-to-face communication, beyond the limitations of space and time, between individuals, communities and generations.

A fourth group of human skills combines memory, the imagination of distant places, future times, and others' needs and desires with the relative valuation of different goods and services and the co-ordination of their supply. This combination leads among other things to the activity of *trade* (prefigured perhaps in the migration and food-gathering of birds). Other species live off the territories they occupy or migrate between, as did humans at first. The realization that human skills and opportunities varied between individuals and places gradually extended trading, from the barter of food for well-made flint hand-axes to ever more complex operations involving money (and its further abstraction as 'finance'), agriculture, craft, manufacturing, cities and transport over increasing distances: this formed the basis of the 'wealth of nations', from ancient Rome and Renaissance Venice to modern New York and Tokyo (see Figure 2.3 overleaf), with all the consequent changes to the face of the Earth. Increasingly, our 'territory' is the whole world, even if we never move from home.

This is not to imply that all such activities have been desirable. Some of the adverse consequences will be considered in Chapter Seven of this book, and in **Taylor (2003)**.

Activity 2.4

Which of the following human attributes is shared with other species, and which is more distinctively human?

(a) the presence of DNA in every cell

(b) using a screwdriver

(c) courtship rituals

(d) predation on other species

(e) paying bills.

Culture, values and lifestyles

The variations among different human societies during the Holocene has been vast, not least in their different responses to, and perceptions of, 'environment'. A simple classification, used for example by Simmons (1989), identifies six 'stages' in human society as:

- proto-humans
- hunters
- early agriculturalists
- advanced agriculturalists
- industrial society
- technological society.

These stages are characterized by ever-increasing use of **extra-somatic energy**,   extra-somatic energy that is energy additional to the somatic energy available in our own bodies from the food we eat (which in turn depends on recent solar energy and photosynthesis, as described in Chapter One). It can take many forms, as we shall see, from draught animals and windmills to coal and electricity. This extra-somatic energy makes possible all kinds of heating and movement – of animals, goods and machines. This development has been seen as 'progress', though

(a)

(c)

(b)

Figure 2.3 The development of trade has been at the root of economic and social development and the resultant globalization: (a) an Ecuadorean village market; (b) the Rialto in Renaissance Venice; and (c) a modern dealing-room.

doubts about the sustainability of modern societies dependent on such high levels of energy use now lead us to be wary of such a view. However, the stages do represent significant *changes* in human lifestyles, accompanied by increasing modification of the environment. Although they developed successively in time, the six stages all still exist somewhere.

Simmons distinguishes the proto-humans, the earlier examples of genus *Homo* (about 2 Ma BP), from the hunters with more varied stone tools and controlled use of fire (about 0.5 Ma BP).The way of life was as **hunter-gatherers** – small family or tribal groups hunting animals and gathering plants, moving about if necessary to avoid using up these resources. This hunter-gathering way of life was the only one – with much regional variation in the balance between the species hunted and gathered – for 90–99 per cent of human existence (depending where you start counting) and is still to be found in isolated areas (for example in the Congo and Amazon). There is conflicting evidence about the extent to which the extinction of the larger animals – **megafauna** – during the last glaciation was due to human hunting or to climate changes. The use of tools and co-operation in hunting and the controlled use of fire certainly gave humans a competitive advantage over other predators. But hunter-gatherer societies have a general need to maintain the species and ecosystems on which they depend, and many have rules and beliefs which try to ensure this (see for example the attitude of the Iroquois to passenger pigeons mentioned by **Hinchliffe and Belshaw, 2003**. While it is difficult to enter into the value systems of vanished societies with no written records, studies of surviving cave-paintings, and sympathetic engagement with the beliefs of remaining indigenous peoples, reveal a rich symbolic life of myths and rituals.

hunter-gatherer

megafauna

The corresponding values might be described as originating in the ethics of the tribe but developing into those of religion: perhaps a belief in the Sun as the essential provider was subsequently embellished with rituals in order to placate it and ensure good harvests. The close identification with the land, plants and animals of the tribe's daily experience and survival engendered attitudes to the ecosystem that are often lost sight of in modern societies.

The major change of the early Holocene was the **domestication** of plants and animals, developing desired properties for food and other uses. Humans became **settled agriculturalists**, largely based in one place with their crops, for example developing rice paddy-fields (see Figure 2.4), or they moved around with their herds as **nomadic pastoralists**, such as the Masai in Africa with their herds of cows (see Figure 2.5).

domestication

settled agriculturalist

nomadic pastoralist

These two patterns – of settled agriculture and nomadic pastoralism – gradually almost entirely displaced hunter-gathering, to form the basis for the rich diversity of lifestyles that characterize most of recorded history until the last few hundred years. They seem to have begun in the Near East (modern Turkey and Iraq), but there were similar developments in Central and East Asia, the Americas and Africa. One chronology for the earliest evidence of domesticated species is shown

Figure 2.4 Paddy-fields developed in China.

Figure 2.5 Masai herdsmen with cattle in a corral formed by traditional housing, Kenya.

in Figure 2.6. The dog pre-dated everything else because it was domesticated as an aid to hunting. Figure 2.6 implies that settled agriculture/pastoralism gradually emerged over a few thousand years in the early Holocene. Conflicts over land use arose between the settled and nomadic groups, with pastoralism usually confined to land less suitable for agriculture. New perceptions of *wealth* arose, with the ownership and exchange of land and/or animals conferring status, power and influence.

An important further transition was from systems for local self-sufficiency to more socially stratified systems with owners separate from labourers, and production

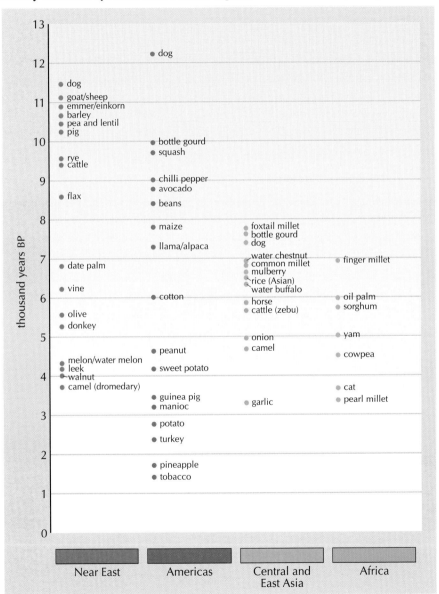

Figure 2.6 Recorded first appearances of individual domesticates in different regions during the Holocene.
Source: Roberts, 1998, Figure 5.4, p.136.

market

Bronze Age shield

for trade as well as personal consumption. In the growing settlements, human craft skills developed as specialized ways of life. **Markets** grew, as trade developed in both agricultural and craft products, with writing and accounting to keep track of them, and their evolving financial structures. The use of bronze, iron and other metals for tools and ornaments involved mining, and the development of chemical processes for their extraction and purification.

Controlled irrigation in otherwise arid regions needed centralized management, inter-group conflicts needed organized armies, public works such as pyramids needed massive impressed labour forces: all these developments contributed to the growth of powerful elites. The ancient empires – Assyrian, Egyptian, Greek, Roman, Chinese, Mogul, Mayan ... – rose and fell, with various forms of totalitarian, feudal or partially democratic government, often based on slavery, and massive waves of migration and conquest. Values now extended from those of the tribe to those of the state, with developing *legal* systems, power *politics* and money *economics*. Religions reflected a new sense of conflict with, or mastery over, Nature, with hierarchies of innumerable gods to be placated. Some of these have given way over recent millennia to single gods – the monotheisms of Judaism, Christianity and Islam – and the non-theistic existential beliefs of Buddhism. The obligations of agricultural and craft routine made human industry

Figure 2.7 Raphael's *School of Athens*: an artist of the Italian Renaissance (400 years BP) recalls the high culture of Ancient Greece (2,400 years BP).

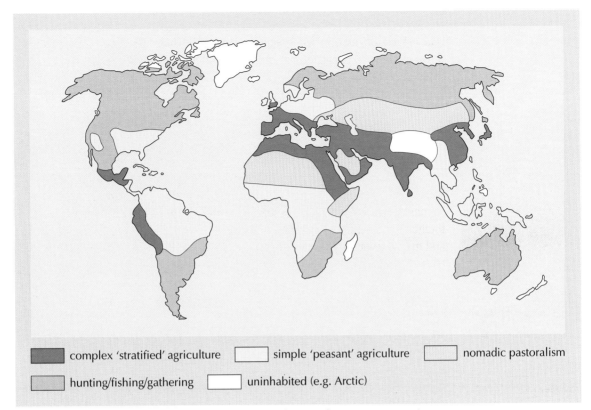

complex 'stratified' agriculture   simple 'peasant' agriculture   nomadic pastoralism

hunting/fishing/gathering   uninhabited (e.g. Arctic)

Figure 2.8 Global cultural developments during the later Holocene (c.2000 BP).
Source: Roberts, 1998, Figure 6.1, p.161.

a necessary virtue, which was arguably less the case in hunter-gatherer societies. However, the wealth created gave leisure to some to cultivate pleasure and philosophical speculation, evolving *aesthetic*, *logical* and *ethical* value systems, some of which are considered in **Belshaw (2003)**: see Figure 2.7. The co-existence of different lifestyles on Earth by about 2,000 years ago is summarized in Figure 2.8.

In substantial areas the world had already been drastically cleared of the vegetation natural to its various biomes. It had been transformed by human activities, particularly in those areas shown in Figure 2.8 as *complex 'stratified' agriculture* with complex social structures, olive groves, irrigated rice fields and other forms of organized land allotment, substantial cities and fortifications. (Virtually the whole landscape of the UK has been transformed by human agency in this way.) From these times we now have innumerable inscriptions and manuscripts to add to the archaeological evidence (see Figure 2.9).

The latest stage, familiar to most of us, is that of **industrialization**, usually located as starting in the UK with the growth of factories, at first water-powered, but increasingly using coal-powered steam engines, beginning as recently as 300 years ago. The process spread rapidly worldwide, with a rapid urbanization of the population (see Figure 2.10) and corresponding changes in the craft, financial and cultural systems.

industrialization

Figure 2.9 Illuminated gospel: manuscript illumination *c.*698/701, Lindisfarne.

Simmons (1989) distinguishes two sub-divisions – *industrial society* and *technological society* – each involving further large increases in energy use, with the later stage introducing electricity and oil-powered transport.

Activity 2.5

What human lifestyles were in evidence on Earth at the following times ?

(a) 10,000 years BP

(b) 2,000 years BP

(c) 200 years BP

(d) 10 years BP.

Figure 2.10 Nineteenth-century industrial city: Gustav Doré's *Over London – by rail.*

A factor in industrialization was the growth of scientific value systems alongside – and often in conflict with – religious ones. This new **scientific method** treats all models and explanations as tentative, only to be retained if supported by experimental evidence, and still always open to revision. This replaced any appeal to written authority, from whatever source – a clear challenge to the sacred texts of some religions. This approach has been manifestly successful in enlarging our understanding of the world over the last three centuries, but its contribution to action in the developing economic climate and political conflicts of our times can lead to problems, as will be considered further in **Bingham et al. (2003)**.

scientific method

An example of this developing scientific understanding is provided by the concept of *energy*, which we have used so far without comment, but is in fact a complex, quite recent way of unifying our view of diverse processes. It will be dwelt on in Section 3 for several reasons: its increasing use distinguishes various lifestyles; the associated skills that will be developed though looking at energy and its measurement will be useful for subsequent chapters; and also because its technological applications are a good example of human-induced change, with the problems that sometimes result.

2.3 Modified ecosystems

The environmental effects of humans can be seen in their impact on ecosystems described in Chapter One. In the absence of human intervention, the only energy source is solar and the chemical elements used are (almost) completely recycled (with some exceptions, like the carbon compounds retained in peat bogs and marine deposits). The species present depend on this access to solar energy and minerals, and there is usually a process of succession to a climax vegetation and fauna when the system has been disturbed (as you will remember from Chapter One). However, in the presence of humans, succession, the relationships between species and all the other processes are much modified. Let us briefly consider these modifications in turn:

- Succession can be drastically interrupted by forest clearance, earth-moving, irrigation, mining and quarrying, growth and harvesting of preferred species and, in urbanization, by the creation and maintenance of almost completely contrived environments.
- While human hunter-gathering is like the predation and foraging of other animals, the interaction with other species has been much modified in later lifestyles (as discussed in **Hinchliffe et al., 2003**). Domestication of plants and animals, and the use of pesticides, seek to limit the community on a farm to a limited range of species (at least of the larger ones). Deforestation, agriculture, industry and urbanization all significantly modify habitats,

denying them to their former inhabitants, with resulting biodiversity changes, as we shall see in Chapter Four.

- Human energy use (though still fundamentally dependent on solar energy to grow food) has dramatically increased with the use of **fossil fuels** (oil, gas and coal) and **nuclear fuels** and these have significant pollution effects on local to global scales.

  fossil fuel

  nuclear fuel

- The range of chemical elements used has likewise expanded vastly from those cycled in natural ecosystems. The traditional stone, bronze and iron of pre-historic tool-making and the ceramics, textiles and metals of historic crafts have been augmented by the ingenuity of modern technology to a bewildering array of lead-acid batteries, quartz-halogen lamps, halocarbon refrigerants, silicon chips and so on. Many of these materials or their by-products create pollutants on their disposal, sometimes producing health or ecosystem damage; they are often difficult to recycle safely, at least on the human timescale.

This goes some way to answering the 'how' of our first question: how and why do human activities change environments? The 'why' is perhaps more open-ended. Obviously, in the first instance humans, like other animals, have a need for food and shelter; success in providing these has supported a growing population, with a corresponding growth in the provision needed (see Figure 2.11). But for at least three millennia human society has developed to provide far more than these primary needs. Landscape is transformed (see Figure 2.12), gardens are created and luxury crops (such as vines, tea, tobacco) grown for pleasure and status. Temples, cathedrals and palaces are built to express religious belief or political power (see Figure 2.13). The rich cultural life of cities is remote from agricultural subsistence (see Figure 2.14). Most of all, in our society, environmental change takes place, often in remote places, because money, and the power it affords, can be made from it.

Figure 2.11 Urban housing: (far left) formal: Fitzroy Street, London; and (left) informal: mud and tin houses, Kenya.

Figure 2.12 American grain prairie, Montana.

Figure 2.13 Krishna Mandir temple in Patan, Kathmandu.

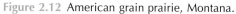

Activity 2.6

Why do *you* think we change environments?

Figure 2.14 Pierre Auguste Renoir, *The Luncheon of the Boating Party.*

Summary

We have seen how distinctively human characteristics have led through the cultural changes of the Holocene to different lifestyles characterized by ever-increasing use of extra-somatic energy, continuing population growth and drastic modification of natural ecosystems. This leads naturally to our second question: what is the nature of energy, and how does its use influence these changes?

3   Energy and society

In everyday life we use the word 'energy', along with related ideas such as force, work, weight, heat and power, in ways that are not always clearly distinguished. But in the developing science of the last few hundred years, these terms have all acquired distinct and narrowly defined meanings with which you should be familiar for the purposes of the discussion in this book. Without attempting a full explanation of all of them, I will try to give the flavour of some general principles, for those who may not be familiar with them, and relate them to human lifestyles and their environmental effects.

3.1   Different forms of energy

○   List some of the different kinds of energy use that you can identify in your daily life.

●   I wonder if you began with the Sun, which is still as crucial as ever? It provides the background light and heating which is all you need in summer (and even in winter is still essential). It also provides, via the ecosystem, the food energy you need for all your physical activities, from climbing stairs to just moving a computer mouse. I expect you thought of the coal, oil or gas used for heating and cooking and, particularly, the petrol or diesel fuel used in cars and other vehicles. Electricity, both from the mains supply and from batteries, provides light and heat, drives electric motors in vacuum cleaners and other machines, and powers refrigerators, microwave cookers, televisions and computers. The mains electricity in turn comes from a coal, gas, nuclear or hydroelectric power station, or even a wind turbine.
And don't forget the cat, which also needs food energy to catch mice, and can warm you by sitting on your lap ...

We can bring some order into this confused list by distinguishing five forms of energy, roughly in the order that they were studied historically, and then by thinking about their conversion from one form to another.

Sir Isaac Newton
(1643–1727)

James Watt
(1736–1819)

John Dalton
(1766–1844)

Michael Faraday
(1791–1867)

Ernest Rutherford
(1871–1937)

The first form is the *mechanical energy* associated with the studies (particularly by Newton in the late seventeenth century) of the way in which objects move in time and space: for example, apples falling to the Earth, planets moving round the Sun, or modern machinery and space vehicles. These studies provided the first way of defining energy so that it could be accurately measured.

The second form is *thermal energy*, the 'heat' that humans have always known about as the warmth of the Sun and that was complemented by the discovery of fire. Watt's refinement of the steam engine, with its crucial role in the Industrial Revolution, stimulated more detailed consideration of its relation to mechanical energy from the eighteenth century onwards.

The third is *chemical energy*, again with a long history of craft knowledge about foods, fuels, explosives, metal-working and so on. This was given a new orderliness in Dalton's atomic theory in the early nineteenth century. At about the same time, apparently separate observations about magnets, batteries, lightning, light and so on were brought together, particularly by Faraday, introducing a fourth form, that of *electromagnetic energy*.

Finally, the twentieth century saw an increasing insight into the internal structure of the atom, bringing the recognition by Rutherford and others of a fifth form, *nuclear energy*, with its dramatic consequences.

To see whether you can identify these five forms of energy, consider this question.

○   What form of energy is contained in each of the following?

a raised weight	a tank of petrol
a coiled watch-spring	a tensed muscle
a rushing stream	a torch battery
a spinning flywheel	the electric mains supply
the sound of an orchestra	the light of the Sun
a pan of boiling water	a dose of X-rays
a bag of coal	a lump of plutonium

●   A short, but rough, answer is that the first five in the first column are mechanical, the next is thermal (boiling water), the last one and the first three in the second column are chemical; the next three are electromagnetic and the last is nuclear.

However, such broad categories may need some explanation and refinement. In order to understand the distinctions, we need to look at the basic principles:

1   Every object has a **mass** (measured in kilograms), which indicates both how   mass
    difficult it is to get it moving ('inertia') and how much it attracts other masses
    ('gravitation').

2   A mass can only be set moving by a **force**, which is also needed to make any   force
    further changes in its speed or direction.

3   When a force moves a mass over a distance, it does **work** (measured in   work
    joules).

4   So finally we get to *energy* (also measured in joules), as the capability to do
    work in this way.

All fourteen of our examples contain energy in this sense, though this may be
more obvious in some cases than others. Energy will be called **mechanical** if it is   mechanical energy
due to the position or movement of mass, and this covers the first five examples –
though they are of different types. Thus the positions of the raised weight and the
coiled spring give them **potential energy**, able to be released to drive different   potential energy
kinds of clock. (While the weight's energy is *gravitational* due to the Earth's
attraction, the spring's is *elastic* due to its internal structure.) The movement of
the stream and the flywheel give them **kinetic energy**, able to turn a millwheel or   kinetic energy
an engine. The sound of the orchestra is perhaps a less obvious example: this is
vibrational energy where the air moves to and fro, able to set a membrane in our   vibrational energy
ears into corresponding motion.

A necessary refinement in all five cases is that each material involved also
contains some **thermal energy**. This is determined by its temperature, and on its   thermal energy
mass and nature. Even though the materials may appear quite cold, they could be
even colder. An object at normal room temperature would, for example, melt a
piece of ice placed in contact with it, so it clearly has some thermal energy.

This is much more obviously the case with the pan of boiling water, which is
therefore a prime example of a thermal energy store (though it also contains
potential energy, if it is raised up on the stove, and kinetic energy, in the
turbulence of the water). The ability to do mechanical work is revealed by its
ability to push up the lid – an observation said to have led to the invention of the
steam engine, which was the first important form of *heat engine* turning heat into
mechanical work.

The bag of coal and tank of petrol can also be used to power a heat engine such as
a steam engine or the internal combustion engine of a car, but they don't do this
because of the potential energy associated with their position or the thermal
energy associated with their temperature. They have **chemical energy** locked up   chemical energy
in their internal structure which is released by the reaction with the oxygen in the
air – the process we call burning. This then turns into the thermal energy needed
to operate the heat engine.

The muscle and battery also depend on stored chemical energy, though again there is some overlap with the other forms. The tensed muscle might also be described as containing (mechanical) potential energy, like the coiled watch-spring – but in this case the energy derives ultimately from the chemical energy in our food. The battery, too, depends on chemical energy in the materials of which it is made (see Figure 2.15). As it gets used, there are associated chemical changes inside it. Because the battery produces useful work when connected to an electrical circuit, it can also be described as containing another kind of energy, in this case electrical. This indicates underlying links between chemical and **electromagnetic energy**.

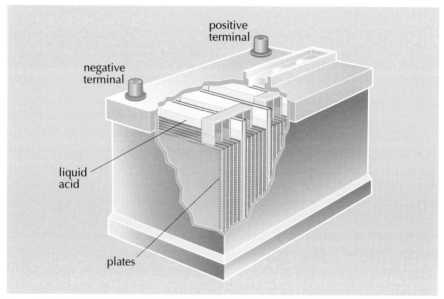

electromagnetic energy

Figure 2.15 A car battery.

The mains supply in your home or workplace also provides electromagnetic energy, though with two differences from the battery. The first is that it is an alternating current, which regularly reverses 50 times per second, a frequency of 50 hertz (Hz). The second is that the conversion from other forms of energy has taken a more complicated route in the power station – perhaps from the chemical energy of coal to thermal energy when it is used to boil water, then to mechanical energy in a steam turbine, and finally to electromagnetic energy in a generator: see Figure 2.16.

What is much less obvious is that the light from the Sun, and the X-rays used in medical diagnostics, also form part of this 'electromagnetic energy' group. These transport energy as **electromagnetic waves,** which can travel outwards from their source – the Sun or an X-ray tube – through empty space or some materials. They are parts of an enormous range of gradually increasing **frequency**, from radio waves at the low end to gamma rays at the high.

electromagnetic wave

frequency

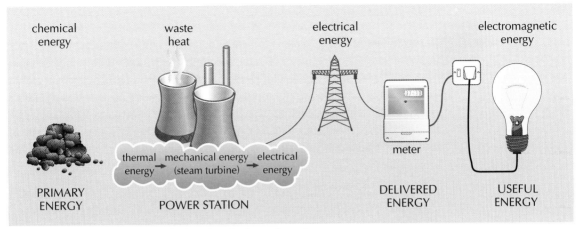

chemical
energy

waste
heat

electrical
energy

electromagnetic
energy

thermal   mechanical energy   electrical
energy   (steam turbine)   energy

meter

PRIMARY
ENERGY

POWER STATION

DELIVERED
ENERGY

USEFUL
ENERGY

Figure 2.16 Energy conversion in the electrical supply system.

Finally, we come to the lump of plutonium. While it also has potential energy associated with its position, thermal energy with its temperature, and its own chemical and electrical properties, its distinctive characteristic is that it provides **nuclear energy** from deep within its structure. This is shown by the fact that it is radioactive, spontaneously changing, and emitting energy that can be used in a bomb, or in the reactor of a nuclear power station.

nuclear energy

Activity 2.7

From the examples given above, complete the following table to summarize the various forms of energy.

Table 2.1

Form of energy	Sub-divisions (if any)	Examples
mechanical	potential	raised weight
chemical		
		sunlight

Compare your answer with the completed table at the end of the chapter.

3.2 Energy conversion

We have already seen how energy can be converted from one form to another – from chemical to mechanical in muscles, from electromagnetic to chemical in photosynthesis, or from chemical-to-thermal-to-mechanical-to-electromagnetic in a power station. The underlying principle is that energy can never be lost; this also means that all forms can be measured in the same unit – the joule. Box 2.2 gives a definition of the joule, the separate unit for **power** (the watt) and the kind of information that can be derived using them.

Box 2.2 Units of energy and power

All forms of energy can be measured in the same unit, the **joule** (J). The primary definition of mechanical energy enables us to use the following expression for the potential energy of a mass relative to some reference level, such as the Earth's surface:

potential energy in joules = mass in kilograms × height above the reference level in metres × 9.81

(Note: the number 9.81 is determined by the Earth's gravitational attraction, so it would be different on another planet.)

○   Newton is reputed to have discovered the principle of gravity when an apple fell on his head. If the apple had a mass 0.1 kg, and it fell on his head from 1 m above, what was its loss of potential energy?

●   Loss of potential energy   =   mass × (initial height – final height) × 9.81
                                 =   0.1 × 1 × 9.81 J
                                 =   0.981 J (i.e. about 1 J)

As a joule is approximately the change of potential energy of an apple between two heights only 1 m apart, it should be clear that it is not a very large unit. So, when considering national or international energy use, very large multiples of the joule are needed, and these have special names in the metric system:

kilojoule (kJ)	= 10^3 J	terajoule (TJ)	= 10^{12}J
megajoule (MJ)	= 10^6J	petajoule (PJ)	= 10^{15}J
gigajoule (GJ)	= 10^9	exajoule (EJ)	= 10^{18}J

Power (in its technical, not its socio-political sense) must be distinguished from energy. Whenever a machine (in the broadest sense, including lamps, living organisms, the power system of a country and the Sun itself) converts energy from one form to another, it does so at a certain rate per unit time. This is its power measured in joules per second, which are called **watts** (W). We are most familiar with this in the 'power ratings' of electrical appliances – a 60 W light bulb converts mains electricity into some light and much more heat (which is unwanted, so the efficiency is low), a 2 kW fire converts mains electricity into mostly radiant heat and a little light (which is unwanted, but the efficiency is high) and so on. The same definition of power applies to other energy converters, such as working horses – say 500 W – and cars – say 50 kW.

watts

In the electrical case, the waters are muddied somewhat by the use of *kilowatt-hour* as a standard unit, for the *energy* converted by a 1 kW device in one hour, which is a rather roundabout way of getting back to joules:

○   How many joules are there in a kilowatt-hour?

●   A 1 kW device converts 1000 J of energy per second from one
    form to another. So in 1 hour (60 × 60 seconds) it converts
    1000 × 60 × 60 J
    = 3 600 000 J = 3.6 MJ, i.e. 1 kWh = 3.6 MJ

Another energy unit which you still come across is the kilocalorie (confusingly called the Calorie in dietary information), which is equivalent to 4.19 kJ.

○   A treadmill at the gym indicates that a 15-minute work-out uses
    87 Calories. What average power rating in watts does this
    represent for the athlete?

●   87 Calories   =   87 × 4.19 × 1000 J
                          = 3.475×10^5 J, used over 15 minutes,
    so average power   = $3.475 \times 10^5 / 15 \times 60$ J per second
                          =   386 W

This is obtained by burning up chemical energy stores in the body, and is a high value, for a brief period of fairly strenuous activity. A more normal **metabolic rate** is from about 100 W when resting to 200 W for more usual levels of activity. A horse has a similar, but higher, range of metabolic power ratings: the imperial unit of the horsepower has now been standardized (optimistically) at 746 W.

metabolic rate

This principle that energy can neither be created nor destroyed is called the **First Law of Thermodynamics**. It means that all conversions can be said to be perfectly efficient, if all forms of energy are taken into account. In practice, however, we are usually interested in only part of the energy obtained from a conversion – the mechanical energy from a heat engine, say, or the food energy from photosynthesis – and regard other forms (very often heat) as wasted. So, practical **efficiency**

First Law of
Thermodynamics

efficiency

means the fraction or percentage, of the total energy supplied that appears in a *wanted* form; the latter is a human judgement, not an absolute property:

$$\text{efficiency (per cent)} = \frac{\text{wanted energy out}}{\text{total energy in}} \times 100$$

As well as conversions between forms, there can also be conversions *within* forms. Thus,

- mechanical potential energy can be converted to kinetic, as in a waterfall;
- kinetic energy of motion in one direction can be converted into that of rotation, as in a windmill;
- kinetic energy can be converted into the vibrational energy of sound by a musical instrument – or by dropping a brick.

Similarly electromagnetic energy can be converted from a battery or the mains
- to radio waves by a radio transmitter
- to light by a lamp
- to X-rays by an X-ray tube

photovoltaic cell

or, in the reverse direction (from high to low frequency) – from sunlight to a current by a solar **photovoltaic cell** (Figure 2.17); such cells are a great hope for future energy policy (and see Figure 2.21).

Electric motors and generators on the other hand involve conversions *between* electromagnetic and mechanical forms: see Figure 2.18.

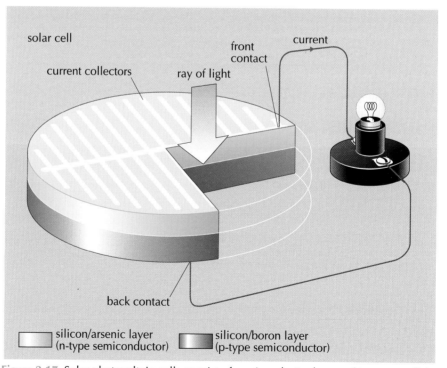

Figure 2.17 Solar photovoltaic cells consist of semiconductor layers of two types: light strikes the junction between the two types, causing electrons to move from the n-type layer to fill holes in the p-type, creating a flow of electrons or an electric current across the junction.

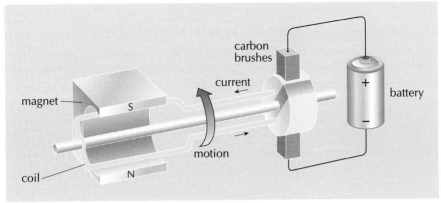

Figure 2.18 A simple direct current electric motor.

Chemical and nuclear energy can be redistributed between different materials by chemical and nuclear reactions, which likewise always involve some other form of energy exchange – thermal, electromagnetic or mechanical. Chemical or nuclear reactions releasing energy, like fuel burning, are called **exothermic**; those requiring energy to be supplied, like photosynthesis, are called **endothermic**.

Thermal energy is rather a special case. Mechanical and electromagnetic conversions are *always* accompanied by thermal effects. In mechanical conversions, there is always some (usually unwanted) heating due to friction. The electromagnetic equivalent of this is resistance to the flow of electric current: this is not always unwanted – the resultant heating is the basis of electric 'fires'. This means that the efficiency of mechanical and electrical machines (except heaters) can never be 100 per cent. But there is a second subtlety about thermal energy, and this is the subject of the **Second Law of Thermodynamics**. One way of expressing this is that thermal energy can only flow spontaneously from hot to cold objects. This happens whenever you cool hot tea with cold milk: though the milk could in theory get even colder, the cold milk does not give up some of its thermal energy to heat up the tea. Put this way it doesn't sound too surprising – but there's nothing in the First Law (energy conservation) to say that this should be impossible.

A practical consequence of the Second Law is that heat engines have a limited efficiency. However much trouble you take to reduce the effects of friction and resistance, their efficiency is proportional to the difference between the initial and final temperatures of the steam or other working fluid. The familiar cooling towers of a power station represent an attempt to make this difference as large as possible – even so, such stations have less than 50 per cent efficiency in converting the energy of the fuel to electromagnetic energy. The rest is lost as 'waste heat' to the atmosphere. If we could find ways to *use* this energy, the efficiency would be much increased. One way is by using it to heat surrounding buildings in a district heating scheme. Such a combined heat and power (CHP) station can then have a total efficiency of 80 per cent or more. These are rare in the UK, but common in Denmark and Germany, for example.

exothermic

endothermic

Second Law of
Thermodynamics

radiation

A final consequence of the Second Law is that all other forms of energy gradually 'degrade' to thermal energy of lower and lower temperature. The same *amount* of thermal energy can be contained in a small mass at a high temperature, or in a large mass at a low temperature, but this represents a different 'quality' of heat. It is lost to its surroundings by **radiation** of different properties depending on the temperature. Thus the solar energy arriving from the Sun (with its very high temperature) is used by life processes and to drive the weather system, all being finally released at the much lower temperature of the Earth, when the same *amount* of energy is lost to space to maintain the Earth's energy balance (otherwise the Earth would heat up; see Chapter Six). As with the heat engine, this change of temperature is necessary to provide energy conversion to other forms.

A summary of these conversion processes within and between forms, with a few not previously mentioned, is included in Table 2.2 for reference – there is no need to memorize the details.

Table 2.2

To From	Mechanical	Electro- magnetic	Thermal	Chemical	Nuclear
Mechanical	potential/ kinetic turbine vibration sound waves turbulence	generator microphone	friction refrigerator heat pump	?	(high- energy collision)
Electro- magnetic	motor loudspeaker	oscillator lamp x-ray solar cell	resistance thermo-electric cooling	electrolysis battery (charge) photosynthesis	high- energy collision
Thermal	heat engine	thermo-electric generation	heat transfer cooling	thermochemistry (endo-)	?(fusion)
Chemical	muscle	battery (discharge) fuel cell photochemistry	thermochemistry (exo-)	+	?
Nuclear	?('radiation' damage)	?(radioactivity)	reactor	?('radiation' effects)	+

Notes: ? means that we do not know how to use these conversions
+ means that these are always accompanied by other conversions.

Activity 2.7

In the following, what forms of energy are converted?

(a) a waterfall

(b) burning wood

(c) a loudspeaker

(d) a nuclear power station.

3.3 Energy in time and space

We have seen how energy use has increased over human history, and how the science of its nature and the technology of its conversion have developed over the past three centuries. We can also recognize that solar energy is converted by photosynthesis and stored for some time in the organisms of the ecosystem before being dissipated as thermal energy in metabolism and decay. Some of these organisms can move over the Earth; those used for human food are often transported over large distances, within and between countries. The concentrated chemical energy in fossil fuels also derives from solar energy stored over the many millions of years required for their formation from plant remains. As fuels are being used incomparably faster than they were formed, a society dependent on them can be said to be living on borrowed time. Many developed countries import fuels and foodstuffs from other parts of the world, so can also be said to be living on borrowed space – an idea developed as the *ecological footprint*, discussed in Chapter Four.

3.4 Fuels and atoms

We saw in Chapter One that there are a hundred or so different elements, each with its own symbol, whose atoms can be combined together to produce the **molecules** of innumerable different compounds. The fossil fuels provide a good starting-point for examining chemical change and its standard chemical shorthand. The burning of carbon (the main constituent of coal) in oxygen (the active constituent of air) to form carbon dioxide (another gas, released into the atmosphere) can be written

$$C + O_2 = CO_2 + energy$$

where C and O_2 are called **reactants**, and CO_2 is a **product**. Each composite symbol represents a molecule. For some elements, like carbon, the molecule and atom are the same (C). For others the stable molecule consists of several atoms – for normal oxygen, two (written O_2). The rarer, but very important, form of oxygen called ozone has three atoms in its molecule (O_3). The compound CO_2 has one atom of carbon and two of oxygen in its molecule. The atoms are held together in molecules by **chemical bonds**, which require energy to break them – and release energy when they are formed. The re-arrangement of these bonds in a chemical reaction provides a net energy change – in exothermic reactions like fuel burning, a *release* of energy.

The burning of natural gas (methane) provides a slightly more complicated example than that of coal:

$$CH_4 + 2O_2 = CO_2 + 2H_2O + energy$$

Here we have molecules of the compounds methane (CH_4) and water (H_2O). At the combustion temperature, the water will be in the form of steam. (A domestic 'condensing boiler' makes sure this is turned back to liquid water to recover heat

molecule

reactant
product

chemical bond

energy that would otherwise be lost to the atmosphere.) This expression also contains 'multipliers' – the 2s to the left of the oxygen and water symbols – that apply to the whole molecule.

○   How many of each kind of atom are there on each side of the equation?

●   One of carbon, four of hydrogen and four of oxygen.

Thus both sides of the equation have the same total number of atoms. This balance is a basic requirement in the equations for all chemical reactions. Atoms are treated as indestructible, with chemical reactions simply rearranging them.

Such equations are not just shorthand. They can include further information, about the masses and volumes of reactants and products and particularly about the amount of energy either released or required when the reaction takes place. For example, the more complex bond rearrangements involved in burning methane are found to give a larger energy release per gram of carbon dioxide produced than the simpler coal (carbon) burning case. This has been used as one argument for using natural gas rather than coal for power generation to reduce possible climate effects (discussed further in Chapter Six).

Not all reactions are exothermic, *releasing* energy in this way. Some, like photosynthesis, are endothermic, requiring energy to be *supplied*. Here the rearranged bonds in the product molecules contain more chemical energy than those of the reactants.

electron

Regularities in the behaviour of atoms, and effects like batteries or photosynthesis, suggest a link between our 'five forms' of energy. Chemical energy can be regarded as a variant of electromagnetic energy, but on the atomic scale. The key idea is that the outer structure of atoms is formed of **electrons**, each of very tiny mass while nearly all of the atomic mass is concentrated in a minute central nucleus. The nucleus and the electrons are kept together by electromagnetic effects. Chemical reactions are concerned with rearrangements or sharing of the outermost electrons, without changing the main structure and identity of the atoms. In some reactions and in solutions and crystals, **ions** are produced. If an atom or group of atoms gains one or more electrons, it is called a **negative ion**, denoted for example by Cl^- or SO_4^-. If one or more electrons are lost, it is called a **positive ion** denoted H^+, Ca^{++} etc.

ion

negative ion

positive ion

Nuclear energy is still seen as a distinct form, originating in changes in the structure of the nucleus itself (so contradicting the idea that atoms are always indestructible). These changes can happen in three ways:

radioactive decay

radioactive isotope

• **Radioactive decay**: this is a spontaneous effect in some naturally occurring atoms, mostly of high atomic mass, and in many artificially produced **radioactive isotopes**. Energetic emissions of alpha particles (atomic nuclei of the gas helium), beta particles (electrons) and gamma rays (electromagnetic waves) convert the elements to others of nearby atomic mass. Each radioactive source has a definite **half-life**, equal to the time taken for half the atoms to decay.   half-life

- **Fission**: this is used as the source of energy in 'atom' bombs and current nuclear power stations, when the nuclei of particular atoms are induced to split approximately in half. These so-called nuclear fuels therefore operate quite differently from chemical fuels, but the resulting energy release can still be used to produce heat to boil water and drive a turbine and generator.   fission

- **Fusion**: this is the source of energy in 'hydrogen' bombs and is also understood to be the source of the Sun's energy, whereby lighter atoms are made to combine together to form heavier ones (e.g. hydrogen to form helium) releasing energy. Controlled use for power stations has proved elusive.   fusion

Activity 2.9

The full 'chemical shorthand' equation for the reaction involved in respiration in living organisms is:

$$C_6H_{12}O_6 + 6O_2 \longrightarrow 6CO_2 + 6H_2O$$

(a) Explain in words what the symbols mean.

(b) Is this an endothermic or exothermic reaction?

(c) What does this mean in terms of the chemical energy contained in the molecules on the two sides of the equation?

3.5 Human energy use

Armed with this more detailed account of energy, its conversion and units, let us now look again at Simmons's classification of human lifestyles: see Figure 2.19 which is taken from Simmons (1989). The first thing to note is that this diagram compares different societies on the basis of their daily per capita energy consumption but measures these in kilocalories (kcal), ranging from 2×10^3 to 230×10^3. So, for our purposes, we need first to standardize them into our preferred energy and power units – joules and watts.

○   What are the highest and lowest daily energy consumption figures in joules?

●   1 kcal = 4.19 kJ, so
2×10^3 kcal $= 8.38 \times 10^3$ kJ $= 8.38$ MJ
Similarly,
230×10^3 kcal $= 963.7$ MJ

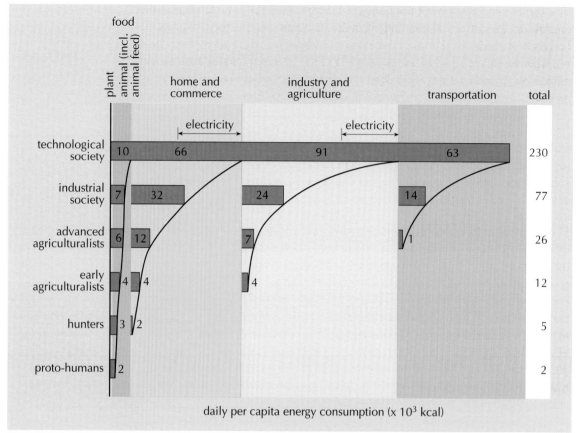

Figure 2.19 The quantities of energy used by human groups at various historical stages and the purposes for which it was used.
Source: Simmons, 1989, Figure 1.3, p.24 (after Bennett, 1976).

○   What do they imply for the average power consumption in watts?

●   1 joule per second = 1 W
So,
$$8.38 \text{ MJ/day} = \frac{(8.38 \times 10^6)}{(60 \times 60 \times 24)} \text{ J/s} = \text{about 100 W}$$

Similarly,
963.7 MJ/day   = about 11,000 W (or 11 kW)

The second thing we might ask is where these detailed figures come from. Indeed, they can only be indicative, depending on the particular societies chosen and the availability of energy data. However, the lowest figure corresponds to the bottom end of the range for human metabolic rate, quoted above as 100–200 W, the primary human energy need. The highest, 11 kW, corresponds well to the rate of energy consumption per capita in a high-tech society like that of the US. European figures are more like 5 kW, while the present *world* average is about 2 kW, with many developing countries well below 1 kW.

The third thing to notice is the way in which the data are broken down. Food energy obtained from plants and animals also includes the food required by domestic animals, so it already includes an extra-somatic component as far as humans are concerned. Extra-somatic energy is also used in home and commerce, in industry and agriculture and in transport, all inexorably increasing over the stages.

Proto-humans obtained their metabolic needs mainly from plants, with limited and sporadic amounts of animal food. Hunters somewhat increased their animal food input, perhaps by more skilled co-operation and the use of fire, but also used some extra-somatic energy, perhaps in the form of food for domesticated dogs, and fuelwood for heating their dwellings.

Early agriculturalists obviously differed in detail between settled and nomadic types. The domesticated animals had to be fed, but provided more animal food in the humans' diet, dung which could be dried and used for heating, and animal power for carrying goods, drawing ploughs and other implements. Some other energy sources were used, such as fuelwood and particularly charcoal for the making of metal tools, not just as an energy source but part of the chemistry of extraction from the ore. All of these processes continue on an increased scale in advanced agricultural societies, with additional use of wind and water power to pump water for drainage and irrigation, and to drive corn-mills. Longer-range transportation of goods and people required more use of horses or other draught animals.

With the move to **industrial society** there was a large increase in energy use, particularly associated with the use of the fossil fuel, coal. Homes and other buildings were heated; factories and railways were powered by steam engines; mining and chemical industries developed. This corresponded to a rapid increase in population, unevenly distributed wealth and air pollution in the growing cities. Agriculture, too, became more mechanized, though the general replacement of horses by tractors belongs more to the final stage, that of **technological society**. This is distinguished by an even larger increase in total energy use, extending the fossil fuels to gas for heating and oil for transport, and the growing provision of electricity as a convenient intermediate 'vector', a means of transferring energy between the primary fuel and the user, for innumerable household, industrial, commercial and agricultural tasks. It was the growing concern that fossil fuels would sooner or later be exhausted that fostered the introduction of nuclear energy as a supplementary form of electricity generation.

industrial society

technological society

Activity 2.10

What rate of human energy use per capita (in kW) is typical of the following?

(a) advanced agricultural society

(b) industrial society.

Summary

This section has examined the nature of energy, recognizing five different forms – mechanical, thermal, electromagnetic, chemical and nuclear – with links between them, according to more recent understanding. Conversion between different forms proceeds without loss, but in practice some forms are wanted, while others (often thermal) are regarded as waste, so efficiency is less than 100 per cent. All forms degrade finally to thermal energy, of progressively lower temperature. We have examined some energy effects in time and space, and the two forms of fuel, chemical and nuclear. Finally, we have re-examined the differential levels of energy used in a variety of human societies, using Simmons's categories.

This begins to answer our second question: what is the nature of energy and how does its use influence these [environmental] changes? We will now look further at the environmental impact of energy use, in order to begin to explore how energy policy might develop in a post-technological society.

4   Impacts of energy use: $I = P \times R \times T$

environmental impact

Energy can be regarded as one of the many resources used by humans in their various lifestyles. As we have seen, it is a rather abstract generalization including such disparate items as fuelwood, flowing water, coal and plutonium. The **environmental impact** of resource use depends both on the scale and the manner of its use. The scale in turn depends on the population using it, and the amount used per head. Together these can be summarized in a notional equation:

$I = P \times R \times T$

or

Impact = Population $\times$ Resource use per head $\times$ Technology used

This is one version of several recent formulations that are designed to emphasize that population on its own is a poor indicator of environmental effects. This is because people in some parts of the world consume far more resources than people in other places. In what follows, I will simply assume that population is a known figure for the world (say, 6,000 M), or a particular region (say, 60 M for the UK), leaving it for the next chapter to discuss the changing and detailed structure of population. The only resource considered here will be energy, with later chapters devoted to the equally important resources of *land*, *water* and *air*, and the economic mechanisms for distributing other resources. I will first consider the scale of energy use (R) in more detail, before reviewing the technology available (T) and its variable environmental impact (I).

4.1 Scale of energy resource use (R)

We have already seen that the energy in various stores can be measured in a common unit, the joule (J), and that the *rate* of its use (power) in various converters (including humans) can be measured in joules per second, i.e. watts (W).

Activity 2.11

Pause for a moment to look back and make sure you understand this distinction.

A broad view of the relationship between energy and power over very wide ranges can be shown in a further pair of logarithmic scales: see Figure 2.20. Make sure you can identify the power ratings quoted earlier – human metabolic rate and electrical appliances – on the lower scale.

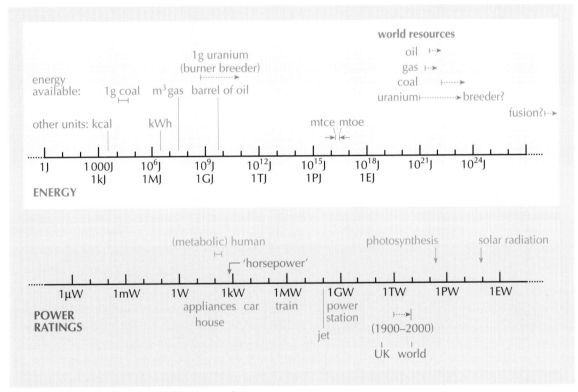

Figure 2.20 Master log scales of energy and power.

The heating requirements of a house can be expressed as an average power rating. If well insulated, this might be about 350 W over the year, but of course this would vary from zero in summer to several kW in winter. Fuel-burning vehicles have powers up to a hundred MW or so for jet engines. Large power stations operate at around 1 GW, while the total average power of the UK energy system is

Biomass is a general term for all the biological materials on Earth that originate in photosynthesis.

around 300 GW, that of the whole world about 12 TW (for commercial forms of energy, with a rather uncertain additional 2 TW or so of traditionally gathered wood and other forms of **biomass**). This figure increased by about 10 times over the twentieth century, though the increase slowed down after the sharp rise in oil price rises in the 1970s, and subsequent concerns about climate change.

Where does all this energy come from? Look at the upper scale in Figure 2.20. At the low end we have the 15 to 35 kJ of chemical energy available from a gram of coal of different qualities; the other fossil fuels – oil and gas – provide slightly larger amounts per gram. Ten thousand to a million times greater is the nuclear energy available from a gram of uranium, depending on how it is used. This massive reduction in the amount of fuel needed has always been seen as a great attraction of nuclear energy. The units of million tonnes of coal equivalent (mtce) and the corresponding million tonnes of oil equivalent (mtoe) are favoured by some large-scale energy analysts. As one million tonnes is 10^{12} grams, these units of 26.4 PJ and 41.9 PJ respectively are 10^{12} times greater than for one gram of the resource.

proved reserve

In the 10^{21} to 10^{23}J range we have various estimates for the world resources of these fossil and nuclear fuels. They start at the **proved reserves** quoted annually, which depend on the current state of exploration and economics, and range upwards to rather vaguer estimates of the total resources available. (The relationship between resources, reserves and technology will be examined further in Chapter Four.) Note that uranium (used in the present way) is a more limited resource than the fossil fuels, and that coal is a much larger resource, also more widely distributed, than oil and gas. However, all such fuels are **non-renewable resources**: that is, they are finite and may eventually all be used up.

non-renewable resource

How do these values compare with world demands? You can judge this by looking across from the power to the energy scales in Figure 2.20: these are not independent of each other, but have been aligned so that the upper one represents the energy converted *annually* at a power rating on the lower one.

○   Can you explain the way in which the scales in Figure 2.20 are aligned?

●   1 W = 1 J/s
There are $60 \times 60 \times 24 \times 365$ seconds in a year
i.e. 3.1536×10^7 seconds.
So a 1 W device operating for a year converts 3.1536×10^7 J, i.e. about 32 MJ (which provides the alignment shown between the two scales).

You will see that the annual world energy demand is not very far below the proved reserves of the fuels – uranium, oil and gas appear able to provide for a few tens of years, coal for a few hundred. Perhaps we shouldn't be surprised at this proximity: these are the *economic* reserves and, as demand increases, more of the total resource becomes economic, as we shall see in Chapter Four. But of course this can't continue indefinitely, and is one reason for constant concern about security of energy supply. Another reason, more evident in recent decades, is that ever-increasing fossil fuel use increases the carbon dioxide concentration in the atmosphere, with implications for climate change: this will be discussed in Chapter Six. With this threat of climate change the question becomes one of whether we can continue to use even the resources we know we have.

Possible ways out of this impasse are also indicated by Figure 2.20. The power scale shows at the high end the continuous rate of arrival of solar radiation on the Earth, some 10,000 times greater than the current world power demand. Even the small fraction of this converted into plant material by photosynthesis provides an *annual* biomass resource comparable to the *total* fossil fuel resources. This solar resource (in its various forms – direct, biomass, wind, wave), along with smaller amounts of tidal and geothermal energy, form the so-called **renewable resources**. They are not as concentrated as the traditional fuel resources,   renewable resource however, so would require major changes in energy policy to utilize properly.

Alternatively, you may note the possible stretching of the uranium resource in so-called **breeder** reactors; these have been widely and expensively investigated but   breeder nowhere are they currently operational. Also note the extremely large theoretical resource for nuclear 'fusion', represented by the hydrogen in sea water, which again has not yet been shown to be feasible or safe. Both of these proposed nuclear solutions fit in with a traditional view of highly concentrated power generation using finite resources, and ever-increasing global energy use. But they involve unsolved problems, and a potential interference with the Earth's energy balance which should not be the case if solar radiation were to be put to better use.

○   If the total world power use is 12 TW and the population is 6,000 M, what is the average rate of energy use per head worldwide, in all forms (i.e. the R in our equation)?

●   12 TW = 12×10^{12} W
6,000 M = 6×10^9 population
so R = 2 kW per head

If the world population stabilizes at 10,000 M as some predictions indicate, *and the average power per head can be kept at 2 kW*, the world power demand will increase to 20 TW, pressing the available resources even harder. However, there are two further difficulties. The first, already mentioned, is that climate change considerations require us to decrease at least our *fossil* fuel use, not increase it. The second is that the 2 kW average power per head conceals a very wide spread between different countries: the USA actually uses more like 12 kW per capita, 4–5 kW is the average use for Europe and Japan, while many developing countries use well below 1 kW per capita. Since at present most of this is fossil fuel use, there is a similar spread in the carbon dioxide emissions. (Something similar applies to all the other resources of technological society.)

Developing countries see it as their right to increase their own use of energy and other resources in order to develop and attempt to match the living standards of the richer nations. In face of this, it seems very unlikely that the 2 kW global average can be maintained and if it were to increase the situation would be even more serious. Thus the P × R product reveals both the pressure on resources (and climate) and the inequity of the wide variation between different countries.

This is leading to some reappraisal of human energy policy, though what a post-technological society would look like on Simmons's energy diagram (Figure 2.20) remains to be seen. Attachment to present policies is very tenacious, as international attempts to negotiate agreed reductions in emissions show.

Activity 2.12

What would be the P × R product for the energy resource use of the following regions?

(a) a developing country with a population of 200 million, and average rate of energy use of 1 kW

(b) a developed country with a population of 50 million, and average rate of energy use of 5 kW

(c) the whole world, with a future population of 8,000 million, and average rate of energy use of 2 kW.

4.2 Varying technology (T)

The T factor – the technology available – is very important, but it has a weakness. Its importance is that it concentrates attention on the way forward out of the difficulties outlined above. Its weakness is that it is much more difficult to assign a numerical value to it, unlike P and, to a lesser extent, R. So consequently I is not a well-defined number either. Most of the controversial aspects of future policy hinge on this less well-defined quantity.

There are two complementary routes. One is to reduce energy demand by being more efficient in the use of energy. At present, heat is lost through poor insulation of buildings, waste heat from power stations is allowed to escape to the atmosphere, inefficient private rather than public transport is preferred, urban design does not minimize transport needs and individual machines of all kinds are less efficient than they could be. Many studies have shown that total energy needs could be halved or more without loss of amenity. Do we really need 5–10 kW for a technological society, or could we converge towards the 2 kW present world average? Of course, with developing countries converging upwards to the same figure, this may not reduce the global total, but it may help to cap it.

The other route is to phase out the supply of energy from fossil fuels, turning instead to nuclear or renewable sources. Nuclear power can be seen as a logical consequence of the view of resource use already embodied in all large power stations: that is, once-and-for-all consumption of a finite resource in large high-tech installations requiring skilled professional management and control, producing difficult-to-manage hazards and wastes. All of these are still the case for nuclear, though it avoids the direct carbon dioxide emissions from all fossil fuels, and the acid rain from coal.

Renewables, mainly solar power in various forms (see Figure 2.21), but also including tidal (arising from gravitational effects) and geothermal (arising from radioactive heating within the Earth), use resources that are continuously available, at least on any conceivable human timescale, without continuously producing wastes. We have already seen that the total solar radiation available greatly exceeds our total needs, though, unlike the concentration of fossil fuels, and even more of nuclear fuels, it is a **distributed resource**.

distributed resource

Figure 2.21   Photovoltaic cells on the roof of the Union Bank of Switzerland administration building near Lugano.

Solar energy can be used directly, either by better design for the way it heats buildings or for water heating on the domestic or power-station scale. It can also be used to generate electricity directly using solar photovoltaic cells based on semiconductor techniques. These are much used in space where cost is not a consideration and are now beginning to be cheap enough for practical use in cladding buildings or roofs. Solar energy can also be used indirectly, by growing biomass such as traditional fuelwood, using straw and other crop residues, or animal and municipal wastes which can be 'digested' to produce methane.

hydroelectric scheme

Solar energy drives the weather system, including the hydrological cycle and hence large **hydroelectric schemes** based on dams (the most established form of renewable energy), or smaller schemes installed in rivers, can be used. Wind turbines for electricity generation are now economically competitive and extraction of wave energy, which arises from the effect of wind over the sea (Chapter Five), is developing. Along with tidal schemes and geothermal installations, there is a wide range of possible options, some of which are, of course, modern versions of methods already in use before fossil fuels (Boyle, 1996). A rough generalization is that where the scale of energy production is comparable to that of power stations – hydroelectric dams, tidal barrages, large wind farms – visual or habitat disturbance can still cause various forms of environmental concern. Very large numbers of small installations may not cause such damage. This suggests that a largely renewable system may require a change to our existing energy, particularly electrical, supply system. We are used to energy *consumption* taking place in innumerable devices, often technically complex (refrigerators, cars, televisions, computers) but designed for ease of use by everyone, whereas energy *supply* is provided by comparatively few, very large power stations. The electricity distribution grid is based on these stations in order to even out supply and demand. A modification of it to permit many more users with their own small generators of various kinds to supply or extract energy at will can be foreseen (if they can be designed for similar ease of use). This seems to offer much greater flexibility in expanding the system, particularly in developing countries, and may be more appropriate to an approach based on renewable resources.

4.3 Varying impact (I)

We turn, finally, to the effects on the environment of energy use in our equation. If we limit our view of environmental impact to greenhouse gas emissions, we *can* be quite specific and assign comparative figures to different forms of energy supply. Such data show that coal is worst, producing about 120 kg of carbon dioxide for each GJ of thermal energy released.

◯   If the world total power of 12 TW were *all* produced from coal, what mass of carbon dioxide would be added to the atmosphere each year?

●   12 TW for one year (32×10^6 seconds) corresponds to
$12 \times 10^{12} \times 32 \times 10^6$ J
$= 384 \times 10^{18}$ J
$= 384 \times 10^9$ GJ
so mass of carbon dioxide produced $= 384 \times 10^9 \times 120$ kg
$=$ about 46×10^9 tonnes.

We can see this as a specific environmental impact, which we might call I_g (for greenhouse impact), with effects on the climate that would need to be determined. Natural gas only produces about 50 kg carbon dioxide per GJ of energy released, so the corresponding I_g (if all energy came from gas) would be about 19×10^9 tonnes of carbon dioxide. Fuel oil falls between coal and gas in its carbon dioxide production.

In this respect, nuclear and renewables both have extremely small emissions. They are not quite zero because there is an energy cost of manufacturing them, which at present is derived from fossil fuel. The appropriate figures are open to dispute, but might be:

- nuclear about 4 kg per GJ
- renewables <1 kg per GJ

So the total I_g for the *greenhouse* impact of the world energy system can be calculated by allowing for the actual mix (the T) of fossil fuels, nuclear and renewables.

Activity 2.13

Of course, greenhouse gas emissions are by no means the only form of environmental impact from energy production. You could make a list of other forms of impact you can think of, to compare with what follows.

Among fossil fuels, coal has the longest history of adverse effects, with the appalling local effects of smoke, soot and smog and waste tips from mining, plus the regional-scale effects of acid rain, particularly from high-sulphur coal used in power stations. Oil likewise produces local pollution from vehicle exhausts, and marine and land pollution from oil spills in extraction and transport. Natural gas (methane) is itself a greenhouse gas much more significant than carbon dioxide, so its leakage is a direct problem. Nuclear power risks major health and ecosystem damage from radioactive releases, and particularly from waste disposal and decommissioning. There are further risks associated with catastrophic accidents, and from weapons proliferation through illicit trade in plutonium. Renewable energy has some environmental effects too: for the larger schemes (hydro, tidal barrages, wind farms) ecosystem disruption or visual disturbance and, for smaller ones, some visual impact, and all have final disposal problems **(Blowers and Elliott, 2003)**. In addition, there are the indirect environmental impacts of different ways of *using* energy in agriculture, industry, housing and transport.

Given the very wide range of effects to be considered, an all-embracing set of T, and hence I, values for different energy supply options is clearly a matter of judgement. For example, we might try to compare the environmental quality of alternative methods by assigning them a T 'index' on a sliding scale from 0 to 1, to be multiplied by P $\times$ R to give a corresponding measure of I. But this would be controversial: how do you weigh global warming against nuclear proliferation, or urban energy supply 'so that the lights don't go out' against a treasured view in the countryside, or the convenience of cars against public transport or home-working? The best that can be said is that fossil fuels, on which we so heavily depend at present, score badly, and that nuclear is not much better or even worse, depending on your strength of feeling about its distinctive problems. Renewables score better, with some further room for debate about the greater impact of large-scale schemes against the more local effects of numerous small devices.

So, in the end, I = P $\times$ R $\times$ T is not so much a calculation as a way of thinking to separate out the issues. Developing countries understandably resent too much emphasis on their growing P while industrialized countries fail to reduce their R $\times$ T, and continue to export methods which will inexorably raise it in other countries. But ultimately, it is the total P $\times$ R $\times$ T (for all resources, not just energy) that will determine total human environmental impact, damage or even destruction.

A final point: this approach implies that 'resource policy' is a matter for rational discussion and planning. In fact the resources used are overwhelmingly determined by what is cheapest in the myriad day-to-day decisions throughout the world, that is, by economic factors under the constant pressure of the immensely powerful supply industries.

Summary

We have seen how the interaction between population, resource use and the technology used can be summarized in the I = P $\times$ R $\times$ T equation, but that attaching numerical values to P and R is much easier than to T, and hence to I. This has been examined in some detail for the energy resource, with the greenhouse impact providing one numerical comparison, but other forms of impact remain more ill-defined. Other resources to be discussed in later chapters will further illustrate the complexity of the problem.

Activity 2.14

What do you understand by the 'equation', I = P $\times$ R $\times$ T?

What does this formulation suggest to us about the relative impacts on our environment of the so-called 'developed' world (Europe/USA etc.) and the 'less developed' world?

5   Conclusion

This chapter has ranged widely over the changes to the Earth induced by human activity over a million years, with a particular emphasis on the social and cultural changes of the last 10,000 years and the large increases in extra-somatic energy use over the last 300 years. Human population has grown faster than exponentially (until recently) and, coupled with changing technology and 'development', so has energy and resource use.

Science has contributed to a better understanding of this energy resource and you should by now understand not only the different forms of energy there are, but also ways of measuring energy output and use. It has also contributed to some questionable technology in the exploitation of energy resources.

We have seen how resource use is very unevenly distributed across the world, but that the consequences of energy use do not respect national boundaries, so it is a global issue that requires an international solution. These issues will be taken up in subsequent chapters, which explore the use of other global resources.

References

Bennett, J.W. (1976) *The Ecological Transition: Cultural Anthropology and Human Adaptation*, Oxford, Pergamon.

Belshaw, C.D. (2003) 'Landscape, wilderness, parks' in Bingham, N. et al. (eds).

Bingham, N., Blowers, A.T. and Belshaw, C.D. (eds) (2003) *Contested Environments*, Chichester, John Wiley & Sons/The Open University (Book 3 in this series).

Blowers, A.T. and Elliott, D.A. (2003) 'Power in the land: conflicts over energy and the environment' in Bingham, N. et al. (eds).

Blowers, A.T. and Smith, S.G. (2003) 'Introducing environmental issues: the environment of an estuary' in Hinchliffe, S.J. et al. (eds).

Boyle, G. (ed.) (1996) *Renewable Energy: Power for a Sustainable Future*, Oxford, Oxford University Press/The Open University.

Freeland, J.R. (2003) 'Are too many species going extinct? Environmental change in time and space' in Hinchliffe, S.J. et al. (eds).

Hinchliffe, S.J. and Belshaw, C.D. (2003) 'Who cares? Values, power and action in environmental contests' in Hinchliffe, S.R. et al. (eds).

Hinchliffe, S.J., Blowers, A.T. and Freeland, J.R. (eds) (2003) *Understanding Environmental Issues*, Chichester, John Wiley & Sons/The Open University (Book 1 in this series).

Ingold, T. (2000) *The Perception of the Environment,* London, Routledge.

Midgley, M. (1979) *Beast and Man: The Roots of Human Nature*, London, Methuen.

Nature (2001) 'Initial sequencing and analysis of the human genome', *Nature*, vol.409, 15 February, pp.860–921.

Roberts, N. (1998) *The Holocene: An Environmental History* (2nd edn), Oxford, Blackwell.

Simmons, I.G. (1989) *Changing the Face of the Earth: Culture, Environment, History*, Oxford, Blackwell.

Taylor, A. (2003) 'Trading with the environment' in Bingham, N. et al. (eds).

Answers to Activities

Activity 2.2

You will have your own answers. I thought of:

- the healing hands of a masseur, a bone-setter; affectionate caresses (perhaps linked to the grooming rituals of other primates);
- the plaiting of hair for adornment; the forming of leaves and straw into containers and roofs; making knots and nets; the shaping of clay into thumb and coil pots, representational forms, walls and building blocks; use of natural earths for bodily adornment and finger painting; assembling delicate machinery;
- athletics, gymnastics, dancing;
- song and speech.

Activity 2.3

There is surely no end to the possible examples!

The hand-tools of traditional crafts: knives, chisels, saws, hammers, screwdrivers, nails and screws ...; spades, hoes, forks ...; knitting needles, crochet hooks ...; scissors, brushes, combs, nail-files ...; rolling pins, baking trays, jelly moulds ...; stirrups, harness ...; daggers, spears, boomerangs ...

and the more elaborate: lathes, milling machines ...; ploughs, wheelbarrows, threshing machines, combine harvesters ...; looms, spinning wheels, the spinning jenny; potter's wheels, kilns, ovens ...; refrigerators, irons, vacuum cleaners ...; guns, bombs, mines, torpedoes, missiles ...; windpumps, mills and turbines, water wheels, power stations ...

those for intellectual or artistic ends: pen and paper, the brush of the artist, the bow of the violinist, the violin itself, other instruments; books, films, tapes and discs, radio, television ...; the typewriter, the computer and the mouse; microscopes, telescopes ...

those for haulage and transport: carts, coaches, lifts, escalators, bicycles, ships, trains, cars, tractors, aircraft, spacecraft ...; robot machines or automatic vehicles ...

Activity 2.4

(a) shared

(b) human

(c) shared

(d) shared

(e) human.

Activity 2.5

(a) hunter-gathering

(b) hunter-gathering and nomadic pastoralism, peasant and stratified agriculture

(c) hunter-gathering and nomadic pastoralism (reducing), peasant and stratified agriculture, and industrial society

(d) hunter-gathering and nomadic pastoralism, peasant and stratified agriculture (reducing), industrial and technological society.

Activity 2.7

My answer is as shown in the following table – but there is some room for variation in the sub-divisions and overlaps:

Table 2.3

Form of energy	Sub-divisions (if any)	Examples
mechanical	potential (gravitational) 　　　　(elastic) 　　　　(chemical) kinetic vibrational	raised weight coiled spring (muscle) stream flywheel sound of orchestra
thermal		(all of above) boiling water
chemical		bag of coal tank of petrol muscle battery
electro-magnetic	steady alternating wave	(battery) mains sunlight X-ray
nuclear		plutonium

Activity 2.8

(a) potential to kinetic (mechanical)

(b) chemical to thermal

(c) electromagnetic to mechanical (vibrational)

(d) nuclear to thermal (to mechanical to electrical in the turbine and generator).

Activity 2.9

(a) As you may remember from Chapter One, the first group of symbols must represent a molecule of carbohydrate, which combines with six molecules of oxygen to form six molecules each of carbon dioxide and water.

(b) Photosynthesis is the reverse of this and is an endothermic reaction, so respiration is exothermic.

(c) Since energy is released, the total chemical energy in the molecules on the left must be greater than in those on the right.

Activity 2.10

(a) 1.3 kW

(b) 4 kW.

Activity 2.12

(a) 200 M $\times$ 1 kW = 200 GW

(b) 50 M $\times$ 5 kW = 250 GW

(c) 8,000 M $\times$ 2kW = 16 TW.

Activity 2.14

The 'equation' suggests that the impact of human activity (I) can be calculated from the total number (P) in the population, the amount and nature of the resources it uses (R) and a factor (T) related to the technology used with those resources.

The population of the developed world is smaller than that of the less developed world, but it uses many more resources per head. At present, many of the technologies used in the developed world also have a high impact, so the developed world has a much higher total impact.

Chapter Three

Population change and environmental change

Michael Drake and Joanna Freeland

Contents

1   Introduction

Populations are key to any environment. If you think of a park, or a city, or a field, it is likely that your picture will be based to some extent on the populations that inhabit that particular environment. For example, you may think of a city as containing populations of humans, sparrows, starlings, rats, ants and dogs, whereas a field in the countryside may contain populations of wheat, poppies, larks, yellowhammers, fieldmice, butterflies and so on. This close relationship between populations and environments means that environmental change can be partly attributed to the ways in which human and non-human populations grow and recede, and exist in various sizes and densities. In order to understand environmental changes it is therefore important to understand populations and population dynamics.

In this chapter we shall be looking at both human and non-human populations. There are several reasons for this. First, given the size and variety of non-human populations, their role in the environment is worthy of attention. Second, although non-human populations are less able than humans to create and mould environments to their will, this is a matter of degree, in that both humans and non-humans have environmental limits. In the case of humans, these are often the products of climate, topography and disease. A glance at a map of human population distribution (Figure 3.1) shows how unevenly we populate the earth, and this is, in part, testament to the environmental limits imposed by various areas of the world. Third, the dynamics of populations – the ways populations rise and fall – evince interesting similarities between humans and non-humans that, in

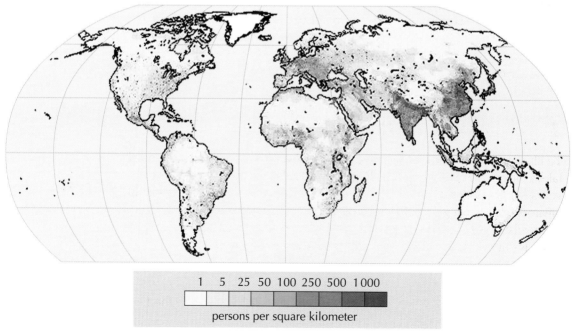

1   5   25 50 100 250 500 1 000

persons per square kilometer

Figure 3.1 Distribution of human populations across the globe.
Source: http://sedac.ciesin.columbia.edu/plue/gpw/index.html

turn, have implications for environmental change. Finally, the interaction between human and non-human populations has, over the centuries, been one of the major determinants of environmental change. Think, for example, of the domestication of certain non-human species and the eradication or decimation of others by humans (see **Freeland, 2003**).

We begin with a brief case study of environmental change and the role population change played in bringing it about. Our purpose is partly to get you to reflect on environmental change in your neighbourhood, and partly to introduce the notion that both the growth and decline of a population may have the *same* environmental impact. Much of this section can be related to the first key question of this book: how and why do environments change?

We then attempt to define a population, before examining the following two related questions: why do populations differ in size, and how do they grow? In answering these questions, we look at human and non-human populations separately, while noting the similarities that seem to unite them. The environmental issue providing the context for this discussion is that of food supply. Here, we are concerned in part with a second key question of the book: how can we represent, model and predict changes and their consequences?

Finally, we come to the sustained period of economic growth since the Industrial Revolution. That growth has been one of the major causes of recent environmental change. We examine the possible role of the size and dynamics of human population in bringing about change and sustaining it. This section of Chapter Three is particularly relevant to a third key question of the book: how do we determine the consequences and significance of changes? However, like the preceding sections, this section has some relevance to all three key questions of this book. As you read through the chapter, bear in mind the key questions, paying particular attention to the overarching question: how and why do environments change?

2   A change in the landscape

To illustrate the environmental impact of the rise and fall of populations, we will take a look at the landscape of the English Midlands over the past one thousand years or so. Changes in the landscape are among the most visual of environmental changes: fields covered by houses, motorways and shopping centres; valleys flooded for reservoirs and hydroelectric stations; and forests cut down for timber or moorlands covered by plantations of conifers.

Activity 3.1

On your next journey to work or to visit a relative or friend, or on the next school run or drive to a shopping centre, make a mental inventory of *three* of the most visually dramatic changes in the landscape on your route that have taken place during the last five years.

In 1300 most of the English Midlands was cultivated under what was called the 'Open Field' system or 'Three Field System'. The population tended to live in nucleated villages surrounded by enormous fields, usually three in number, hence the term Three Field System. These fields might extend to several hundred hectares each. (To give you a sense of scale, a football pitch is about 2.5 acres or 1 hectare.) These fields were cultivated by the inhabitants of the village, with each occupier having a portion of the fields, often consisting of several different parts scattered across each field in strips. These parcels of land within the large fields were not bounded by hedges or fences – they were 'open' – and because of this a common production policy had to be followed. Thus, in the first year, one of the large fields would be sown in the autumn with wheat or rye, a crop that all those occupying land in that particular field were obliged to sow, cultivate and harvest according to a common timetable. The wheat or rye would be harvested in year two. In year three, barley or oats would be sown in the spring. In the fourth year the field might be left fallow, in an attempt to restore its fertility. By 1300 this system had already existed for several hundred years and was to last virtually unchallenged for another 500, as you can see from Figure 3.2(a).

However, some changes were being made. When new land was taken into cultivation as a result of clearing woodland or draining marshes, it might not be added to the open fields; rather, it was hedged or fenced and cultivated as a separate entity. In addition, the Black Death (1348–1352) and further outbreaks of bubonic plague brought a dramatic fall in the population of England and Wales: it is estimated that it fell from about 5.5 to 6.5 million in 1300 to about 2.0 to 2.5 million in 1400, not returning to its medieval peak until the eighteenth century. Demand for grain fell and less favourable land was turned over to other uses or abandoned. Hundreds of villages and hamlets were deserted and urban populations shrank. The fall in the rural population eased the rigidities of the common production policy of the early medieval system of agriculture. There was, too, some increase in the demand for English wool, especially in overseas markets, so more sheep were kept. All this encouraged the idea of dividing up the open fields, or parts of them, into discrete entities: a development known as **enclosure** (see Box 3.1). A hedged field was particularly valuable for stock raising, for not only did it make it easier to control grazing animals, but the hedges also provided shelter for the animals. That is why, until recently, hedges – especially in pastoral or mixed farming areas – were both high and thick. Enclosure was carried out on a piecemeal basis over a long period of time (several centuries) on a more or less voluntary basis – certainly it was not backed by law as were the later Parliamentary enclosures. It led to the patchwork pattern of relatively small fields, a familiar feature of the English Midlands until half a century ago that is still characteristic of parts of the West Midlands and illustrated in Figure 3.2(b) and (c)

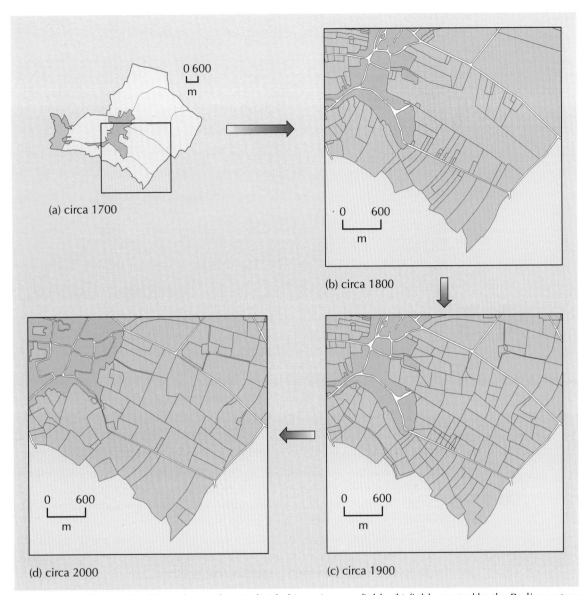

(a) circa 1700

(b) circa 1800

(d) circa 2000

(c) circa 1900

Figure 3.2 (a) The parish of Raunds (Northants) divided into six open fields; (b) fields created by the Parliamentary enclosure of Raunds in 1798; (c) part of the field system in Raunds (Northants) at the end of the nineteenth century (note the creation of yet more fields after 1798); and (d) the twentieth century saw the move back to larger fields. *Source:* adapted from Hall et al., 1998.

Although this voluntary enclosure only nibbled at the Three Field System in its heartland of the English Midlands, it did have a dramatic impact on the landscape in the areas where it occurred. 'By 1700 more than two-thirds of England was enclosed by hedges and walls, and a quarter of the country had been enclosed in

Figure 3.3 Aerial photograph of part of the field system of Raunds (Northants) at the end of the twentieth century. Note the somewhat vestigial nature of some of the hedges. *Source:* www.multimap.com

Box 3.1 Parliamentary enclosure

Parliamentary enclosure (circa 1750–1850) is so-called because each set of open fields needed an Act of Parliament to trigger the enclosure procedure. Previously, any enclosure that occurred had been on a voluntary basis, with groups of landowners re-allocating the bits of land they owned in the open fields into compact units, which they then fenced or put a hedge around. This voluntary system worked where the land was owned by relatively few people. In areas where there were large numbers of landowners, getting the agreement of all proved very difficult. Hence the recourse to Parliament, which allowed the enclosure to take place when the owners of two-thirds of the land agreed.

the course of the seventeenth century' (Overton, 1996, p.148, quoted in Slack, 2002). The Parliamentary enclosure movement had the same effect. Even a casual glance at Figure 3.2 (a) and (b) will bring this out, and the enclosures did not end with the Acts of Parliament (see (b) and (c) in Figure 3.2). Note the growth in the number of enclosed fields in Raunds in the course of the nineteenth century.

The mid-eighteenth century saw the beginning of the end of the Three Field System. In the course of the next hundred years it virtually disappeared, to be replaced by a patchwork of relatively small fields. Although the environmental impact then was the same – at least visually – as the previous piecemeal and voluntary enclosure system, the effect of population could not have been more different. Whereas the previous enclosure movement was a response to falling population, this time population increase was partly the cause. The population in England and Wales during the hundred years 1750–1850 rose to ever higher peaks, from 5.8 million in 1750 to 8.7 million in 1800 to 16.7 million in 1850 (Anderson, 1996, pp.77, 211). This led to a rise in the demand for grain which led to higher prices, especially during the French Revolutionary and Napoleonic Wars (1793–1815), when England was cut off from overseas supplies for long periods. The stimulus of rising prices led to a demand for enclosure as this allowed farmers and landowners to adopt new techniques, new strains of grain, and new rotations of crops, involving roots such as turnips and grasses, without – as they saw it – being held back by the imperatives of the Open Field System, with its communal restrictions on land use. Because the new rotations involved crops other than grains, it was now less necessary to let the fields lie fallow every so many years in order to restore their fertility. This helped increase the productivity of the land and changed the landscape further. Enclosure by Act of Parliament, with its element of compulsion, replaced voluntary enclosure.

Taken together, the voluntary and Parliamentary enclosure movements took several hundred years to change the landscape of the English Midlands. It has, however, taken only 60 or so years (from 1940 onwards) for it to start to move back decisively towards large fields (Figure 3.2(d) and Figure 3.3), reminiscent of the medieval agricultural system. Since the start of the Second World War the demand for the home production of foodstuffs has increased dramatically, but not because the population has grown particularly rapidly – it hasn't! Rather, the impetus has come from governments anxious to be self-sufficient in agricultural products. The move back to large fields, which has entailed the removal of tens of thousands of miles of hedges so painstakingly planted and maintained (for half a millennium in some cases), has been a product of three factors: the new technology – especially machinery – the use of which was inhibited by the patchwork of small fields created by the enclosure movement; alternative methods of controlling animals, such as electric fencing; and the cost of maintaining traditional hedges.

○   What do you think was the impact of enclosures on non-human populations?

●   Although, as we shall see later in this chapter, the growth and expansion of human populations has been largely at the expense of non-human populations, especially those living in the wild, the development of hedged fields was one step in the opposite direction. Hedges, by providing shelter and food, greatly increased the habitats of certain non-human species – for example, birds that nest and/or roost in hedges, and mice, hares, rabbits and so on, that shelter and breed there. In addition, many plant species, together with a wide diversity of insects and other invertebrates, can be found thriving in hedgerows.

Although, so far as the English countryside is concerned, a sharp fall and then a sharp rise in the population contributed to the development of the *same* mosaic of English fields, this change in the visual environment was associated with quite marked differences in the social and economic environment. For instance, although the landless labourer was not uncommon in the Middle Ages, the breakdown of the Open Field system was accompanied by changes in the social make-up of the countryside, with a shift towards a triumvirate of landowners, tenant farmers and landless labourers. The last named began to fall in numbers in the second half of the nineteenth century and have now all but disappeared. Within recent decades too, the number of farmers has also declined dramatically.

The apparently enduring – and endearing! – appearance of much of the English countryside over recent centuries, has been associated with enormous changes in the economic environment too, with a shift from primary (agriculture, forestry, farming, mining) to secondary (industry) to tertiary (services) activities as the focus of employment and national income. This has been accompanied by heavy urbanization, so that as early as 1900 some 80 per cent of England's population lived in towns and cities.

Summary

In this section we have used changes in the English rural landscape, more particularly the size of fields, to illustrate the relationship between population change and environmental change.

The dramatic fall in England's population in the late fourteenth century and early fifteenth centuries, a fall that was not made good until the early eighteenth century, was associated with the tentative beginnings of a move from the so-called Open Fields to the patchwork of relatively small fields that was such a feature of much of the English landscape until the second half of the twentieth century. We then saw that the Open Field system effectively disappeared during the period 1750–1850, which also happened to be a period of unprecedented population growth – a trebling of it. Since the 1940s there has been a move back towards the large fields of the Middle Ages, although during this time (1940–2000) the country's population has risen by only 20 per cent.

The somewhat crude message of this story is that there is no simple correlation between population change and environmental change. We have seen that over the last 600–700 years there have been dramatic changes in English agriculture, in terms of crops, yields, farming practices, technological changes and social relations. The overall economic environment has also undergone enormous changes, the population now about ten times as big as at its medieval peak, but with agriculture employing less than one tenth of the population it did *then*. Nevertheless, as we will find in the next section, populations do matter even if their size is not the only significant feature.

3   What is a population and why do populations differ in size?

Population can be defined as a group of individuals from the same species that occupies a geographically restricted area. For some species, a population can be further defined as containing individuals that have the potential to breed with one another, although this is relevant only to species that sexually reproduce.

population

3.1 Variations in size and density

The number of individuals within populations (population size) varies enormously. For example, a colony of ants in a back garden may represent a population of more than 1,000 individuals (depending on the age of the colony), whereas the population of grizzly bears in Ivvavik National Park in Canada has no more than 150 individuals. *Within* the species we also get enormous variations in the number of individuals in a particular population. Today, for instance, the human inhabitants of the city of Tokyo number 26.4 million and although that is the most populous city in the world there are many others containing populations many times larger than those of whole countries (Table 3.1). When looking at these figures you should bear in mind the difficulties in counting populations: many people are 'unofficially' present in some cities and this will add to the number in the population of such cities.

The numbers in the population data in Table 3.1 are very different from, say, the numbers of bushmen in the enormous expanse of the Kalahari Desert, who total only a few thousand.

Table 3.1   Megacities of the world: cities with 10 million or more inhabitants in 1975, 2000 and 2015 (projected)

1975		2000		2015 (projected)	
City	**Population (in millions)**	**City**	**Population (in millions)**	**City**	**Population (in millions)**
		*London	7.5	*London	8.2
Tokyo	19.8	Tokyo	26.4	Tokyo	26.4
New York	15.9	Mexico City	18.1	Bombay	26.1
Shanghai	11.4	Bombay	18.1	Lagos	23.2
Mexico City	11.2	Sâo Paulo	17.8	Dhaki	21.1
Sâo Paulo	10.0	Shanghai	17.0	Sâo Paulo	20.4
		New York	16.6	Karachi	19.2
		Lagos	13.4	Mexico City	19.2
		Los Angeles	13.1	Shanghai	19.1
		Calcutta	12.9	New York	17.4
		Buenos Aires	12.6	Jakarta	17.3
		Dhaki	12.3	Calcutta	17.3
		Karachi	11.8	Delhi	16.8
		Delhi	11.7	Metro Manila	14.8
		Jakarta	11.0	Los Angeles	14.1
		Osaka	11.0	Buenos Aires	14.1
		Metro Manila	10.9	Cairo	13.8
		Rio de Janeiro	10.8	Istanbul	12.5
		Cairo	10.6	Beijing	12.3
				Rio de Janeiro	11.9
				Osaka	11.0
				Tianjin	10.7
				Hyderabad	10.5
				Bangkok	10.1
Total	68.3		263.6		387.5

* London's population included for comparison.

Source: Population Information Program, 2000, March.

The world urban population (the definition of 'urban' used here is a settlement with a minimum population of 2,000) is currently as high as three billion – half the world's total population (see Figure 3.4) – having risen from 750 million in 1950. It is expected to reach 5 billion by 2030 (Johns Hopkins Bloomberg School of Public Health, 2000, p.12).

○ Look at Figure 3.4. What strikes you about the rate of population increase over time?

● The rate of increase has been accelerating over time, with the rise in population between 1950 and 2000 amounting to, say, seven times the total population of the world in the mid-seventeenth century.

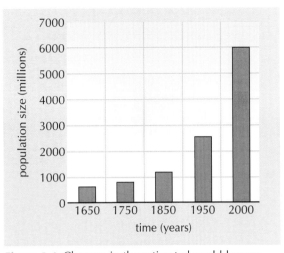

Figure 3.4 Changes in the estimated world human population size 1650–2000.
Source: data from Johns Hopkins Bloomberg School of Public Health, 2000, p.12.

As with humans, non-human populations occupy very different areas of land: somewhere in the region of 0.3 m^2 for the population of ants we mentioned above, compared to 1,000 km^2 for the population of bears. Information on the size of a population and the area that it occupies can be used to calculate the **density** of any given population. In these examples, the ant population includes a large number of individuals living within a relatively small area (this could be somewhere in the order of 1,000 individuals/m^2), which means the ant density is relatively high (Figure 3.5). In contrast, the grizzly bear population includes few individuals in a large area (150 bears/1,000 km^2 which, to put it on the same scale as the ants, is equivalent to 0.0015 individuals/m^2; MacHutchon, 1996), which means that the grizzly bear density is relatively low. To return to our human analogy, the density of the population of Tokyo in the year 2000 was evidently vastly greater than that of the bushmen of the Kalahari Desert.

Figure 3.5 Red garden ants (*Myrmica rubra*) eating a caterpillar.

3.2 Changes in population size

Variations in population size and density require some explanation. Generally speaking, the balance of four processes determines whether a population size increases, decreases or remains constant. These processes are: birth, death, immigration into the population and emigration from the population. Death and emigration reduce population size, whereas birth and immigration increase population size. At any point in time, populations are difficult to measure and their change over time even more difficult to forecast, one reason being that they are in constant flux.

To get a feel for the problem, imagine for a moment a box of different coloured marbles. It has a hole near the top and an outlet at the bottom. Marbles are fed into the box at a certain rate through the former and released from it at a certain rate through the latter. At any particular time the size and structure of the population of marbles (how many marbles of one colour, how many of another) will depend upon the size of these two rates and the length of time marbles have been retained in the box. Let us suppose that these rates are exactly equal at the present time (not a common occurrence in the case of populations of living creatures): thirty marbles per minute coming in, thirty marbles per minute going out. The number of marbles in the box will remain the same. But the composition of the body – or population – of marbles will be changing constantly. For every minute, thirty 'new' marbles are coming in and thirty 'old' marbles are going out. The analogy with human and non-human populations is pretty obvious. If we think of the marbles coming in as births and immigrants and those going out as deaths and emigrants, with the marbles in the box as a population, we can see that one way of defining population change is as *a flow of individuals through time*.

A population is, therefore, a dynamic, not a static phenomenon, and it is this particular characteristic of populations that poses special problems for anyone trying to measure population change, and even more so when they try to predict its future behaviour. Because populations are changing constantly and because their structure affects the processes by which they change and these processes in turn affect the structure of the population, the problems of measuring and of ascertaining the nature of these interactions is very complex indeed. For instance, with respect to human populations, if in a given population the number of females in the fertile ages rises, there is a tendency for the birth rate to rise. Whether it does so or not will depend on a variety of social and economic factors. Similarly, the number of deaths in a population will rise if the proportion over, say, 70 years of age rises. High birth rates will tend to lead to a population with a high proportion of children; low birth rates a low proportion of children. **Population pyramids**, which represent the structure of popula-tions pictorially, have, then, many different shapes. Pre-industrial

Population pyramids illustrated below: (*upper*) Egyptian pyramidal shape, Singapore, 1969 (estimated); (*lower*) tubular shape, Scotland, 1969 (estimated).

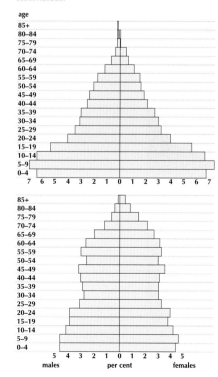

societies with high birth and death rates tend to produce population pyramids looking much like the pyramids of ancient Egypt, with a wide base, representing the younger ages, tapering to a point at the highest age. Post-industrial populations have pyramids with a more tubular shape as their birth and death rates are comparatively low.

The carrying capacity of populations

Populations can fluctuate over time, but there is an upper limit to the size of population that a particular area can support because there are generally finite supplies of food, water, shelter and other resources. This upper limit is called the **carrying capacity**, also designated K, and represents a population's maximum stable size (the number of individuals that can be supported within a population from one year to the next). The carrying capacity of animal populations is generally determined by factors such as food availability, number of breeding sites, adequate shelter or any other resource that may be in limited supply. The different range sizes of grizzly bears give us some idea of how carrying capacity can vary with the environment. Range size is the area that an animal covers in its annual quest for food, shelter, mates, etc., and in grizzly bears it varies from 1,350 km^2 for an adult male in the Brooks Range in Alaska to 27 km^2 in salmon-rich coastal areas in Alaska and British Columbia. The main difference between these two areas is food availability. Salmon provide an ample and nutritious food that reduces the need for wide-scale foraging by the bears (Figure 3.6). In contrast, bears living inland often have to survive on a varied, generally poorer diet that includes nuts and berries that must be collected over a wide area. In other words, the carrying capacity for bear populations in salmon-rich areas is higher than for populations inland because the salmon-rich areas provide much more food per square kilometre.

carrying capacity (K)

Figure 3.6 Grizzly bear (*Ursus arctos horribilis*) fishing, Alaska, USA.

Figure 3.7 Predator–prey relationship: a Canadian lynx (*Lynx canadensis*) chasing a snowshoe hare (*Lepus americanus*).

In the example of grizzly bears we learned that food is an important regulator of population density – and therefore total population size. The relationship between food availability and population size can be further examined by looking at predator–prey relationships. Probably the best known example of predator–prey relationships is that of the Canadian lynx (*Lynx canadensis*) and snowshoe hare (*Lepus americanus*), which is illustrated in Figure 3.7. The snowshoe hare is the major food supply for lynx in Canada, and pelt-trading records of the Hudson Bay Company collected over almost a century have provided a good record of how population sizes of these two species fluctuated over time. Changes in population sizes over a ninety-year period are shown in Figure 3.8.

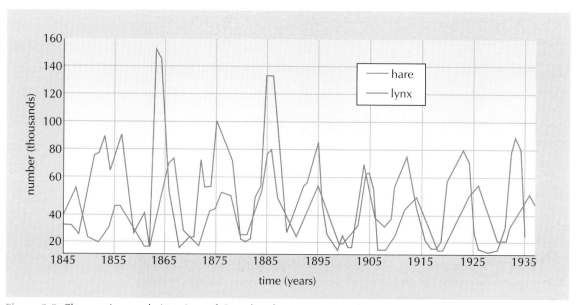

Figure 3.8 Changes in population sizes of Canadian lynx (*Lynx canadensis*) and snowshoe hare (*Lepus americanus*) between 1845 and 1935.
Source: www.forestry.auburn,edu/ditchkoff/images/lecture%20Images/Carnivores/lynx-hare-cycle.gif, redrawn from MacLulich, 1937.

Activity 3.2

Take a look at Figure 3.8. Does there appear to be any relationship between hare population sizes and lynx population sizes?

Comment

The cycles of the two populations are roughly synchronous: when the lynx population is large, the hare population is large, and vice versa.

We deliberately described the lynx and hare populations as being only *roughly* synchronous because within each cycle the maximum sizes of the lynx population often occurred slightly later than the maximum sizes of the snowshoe hare population. We can make sense of this by first imagining a time in which the hare population is growing. As the hare population size increases, there will be more food for the lynx, and therefore the lynx population will also increase. However, as the lynx population grows, more and more hares are being eaten, and the hare population begins to decline as a result of this predation. Food then becomes a limiting factor for the lynx, and its population will then start to decline. As lynx numbers go down, predation decreases, the hare population starts to increase, and the cycle is repeated. In this case, food (hares) is determining the carrying capacity of the lynx population. The predator–prey cycle is summarized in Figure 3.9.

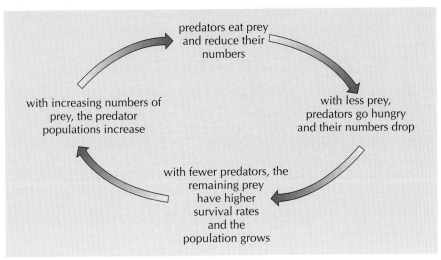

Figure 3.9 How populations may cycle over time.

Activity 3.3

What might happen if the predator (in this case the lynx) was removed from the predator–prey relationship? Can you think of any examples in which prey populations have flourished in the absence of predators?

Comment

The prey (in this case the snowshoe hare) will remain unchecked and populations will continue to grow. Presumably at some point they will be limited by resource availability, but the population may get very large before this happens. There are many examples of this, such as deer and rabbits in the many countries that no longer have enough natural predators to keep their populations in check.

As we shall see, populations in the wild are seldom regulated in a straight-forward manner; even in the lynx–snowshoe hare example above, factors other than predation, such as climate and disease, are involved in population regulation. Nevertheless, predator–prey relationships and, similarly, plant–herbivore relationships are widespread in nature: for example, lions and gazelles, birds and insects, pandas and bamboo trees, red grouse and parasitic nematode worms (Figure 3.10 illustrates some other examples). Although predation and food supply are only two of the factors that regulate populations, their roles are extremely important in the checks and balances of population growth.

(a)

(b)

(c)

Figure 3.10 Herbivore–plant and predator–prey relationships are wide-ranging in nature: (a) koala (*Phascolarctos cinereus*) eating eucalyptus leaves (*Eucalyptus* sp.), (b) garden spider (*Araneus diademetus*) with fly prey wrapped in silk, and (c) European skipper (*Thymelicus lineola*) feeding on Viper's Bugloss (*Echium vulgare*).

As we have seen, human populations differ in size too, but this is largely because of differences in technology, with the carrying capacity for human populations being 'a function of the kind of productive technology which is used' (Wilkinson, 1973, p.16). This reflects the assertion that 'the evolution of cultural systems replaced biological evolution as man's [sic] most important means of making adaptive changes' (Wilkinson, 1973, p.10). The relationship between viruses/bacteria and human populations demonstrates, however, that human populations have never been masters of their own fate – neither in the past nor in the present. For instance, the bubonic plague reduced by half the population of Europe in a matter of a few years, between 1348 and 1352 (**Blowers and Smith, 2003**). Repeated epidemics of the disease kept the population down for many years: England, for instance, did not recover its population size of 1300 until 1700. The well known influenza epidemic immediately after the First World War also wreaked havoc. More recently, AIDS is expected to account for one third of all deaths in South Africa in the year 2001, with two-thirds of all deaths being AIDS-related by 2010 (*The Independent*, 17 October 2001).

Summary

There are many reasons why populations differ in size, but in this section we have been particularly concerned with exploring the relationship between the availability of a particular resource – food – and total population size. Food is often a major determinant of a population's carrying capacity, K. Carrying capacity determines the maximum density of a population which, within a specified area, determines the total population size. Fluctuations in the population size of one species often impact upon the population sizes of other species.

4   How do populations grow?

Populations seldom grow in a straightforward manner. In this section we look at the different stages of population growth. We also learn how the rates at which populations grow allow us to place most species into one of two categories: rapid growth and long-term instability, or slow growth and long-term stability.

4.1 Non-human populations

Let us imagine for a minute what will happen over a 5-year period to a population of deer in which the birth and death rate remain constant and there is no immigration or emigration (a drastically oversimplified deer population!). If every 100 adults produced 200 fawns each summer, and a total of 50 individuals in the population died, there would be an increase each year of 150 new deer for every 100 deer, an increase of 150 per cent (or 150/100). This value of 150 per cent is a *relative* increase because it describes the increase relative to the number of individuals that already existed. If, in the following year, there was the same percentage (or relative) increase, we would have **exponential population growth (Freeland, 2003)**, with a constant relative growth rate of 150 per cent per year.

exponential population growth

However, while the relative increase is constant from one year to the next, the actual or *absolute* change in numbers (that is, the difference between the number present at the beginning and end of each year) is not constant: if population growth remains exponential the absolute increase will be larger in each subsequent year. If we start with 100 adults in year 1, then by year 2 we will have 250 adults (a difference of 150). These will produce 250 x 150/100 fawns, so in year 3 we will have 625 adults (a difference of 375) and so on. Each year sees an increasing number of new adults, and the number by which the population increases each year gets bigger and bigger. This exponential growth rate is shown by stages 1 to 2 in Figure 3.11.

Activity 3.4

If the carrying capacity is not reached, and the growth rate remains constant, what will be the increase in the size of our deer population in years 4 and 5?

Comment

Year 4: 625 x 150/100 = 937.5 (round up to 938)

Year 5: 938 x 150/100 = 1407

As mentioned above, stages 1 to 2 of Figure 3.11 show typical patterns of population increase under exponential growth. The line in stage 1 is nearly flat, which means that the population number has not changed very much over this period. In contrast, the line through stage 2 is very steep, which reflects a much higher increase in population size over this time.

○   In stages 1 and 2 of Figure 3.11, which is more likely to be constant: relative growth rate or absolute growth rate?

●   Although stages 1 and 2 show very different growth rates, the relative growth rate may be unchanging, whereas the absolute growth rate is clearly very different.

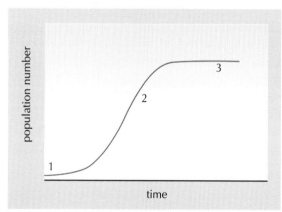

Figure 3.11 A general example of population growth. Stages 1 and 2 represent growth stages, and stage 3 represents the carrying capacity when the population size remains relatively constant provided sufficient resources are maintained.

r
r-species

In Figure 3.11 we have also introduced a third stage of population growth – stage 3 is the plateau that occurs when the carrying capacity is reached. As mentioned above, our deer population example was drastically oversimplified because we assumed, among other things, that it would continue to grow indefinitely. In reality, population growth would slow and eventually stop when the population reached its carrying capacity.

The rate at which a population grows is usually designated as *r*. Patterns of population growth can be placed into one of two categories. In the first category, think of species such as weeds or aphids in your garden, which can cause infestations because of their ability to reproduce very quickly. The reason that populations of these species can grow so rapidly is that individuals start to reproduce at an early age and have a large number of offspring. Species within this first category are known as *r*-**species**, which reproduce by producing as many offspring as soon and as often as possible (Table 3.2) in the hope that some will survive. The cost of this strategy is a high mortality rate – many offspring will die as they are not well provided for. *r*-species are commonly found in unstable or temporary environments such as during the early stages of succession (see Chapter One), and their early and rapid onset of reproduction means that they are often able to produce offspring before the environment changes and becomes unfavourable for them. Populations of *r*-species have a tendency to regularly increase and decrease in size.

The second category of population growth, the so-called **K-species**, includes K-species birds and mammals (including humans). *K*-species tend to live in stable environments and their populations are often maintained for relatively long periods at their carrying capacity, or *K* (hence the term *K*-species). An example of this is oak trees in woodlands: if undisturbed, they may live for several hundred years, although if the forest is destroyed it will take the oak population many years to re-establish itself. Fluctuations in population size are generally less dramatic in *K*- compared to *r*-species. Characteristically, age of first reproduction in *K* species is later than in *r*-species and, in the case of animals, more investment is placed in each offspring both in terms of development (offspring are relatively large) and care (which includes providing offspring with food for a period of time after they are born). Table 3.2 summarizes some of the main differences between *r*-species and *K*-species. It is worth noting at this point that the differentiation between *r*- and *K*-species is not always clear-cut, but this comparison is a useful starting point for models that are concerned with predicting changes in populations.

Table 3.2   Some common characteristics of *r*-species and *K*-species

Feature	K-species	r-species
number of offspring per breeding cycle	few	many
size of offspring	large	small
size of adult	large	small
life span	long	short
Examples	ducks, deer, oak trees	weeds, aphids, spiders, fish

○ Using the information provided in Table 3.2, in which category (*r* or *K*) would you place grizzly bears and ants?

● Grizzly bears are *K*-species; ants are *r*-species.

To make sure that you understand the principles of population growth, turn now to Activity 3.5 and Figure 3.12.

Activity 3.5

On Figure 3.12, mark the apparent carrying capacities and periods of growth for each population, and then describe how the values of *r* and *K* compare between the two populations in Figure 3.12 (refer back to Figure 3.11 if necessary).

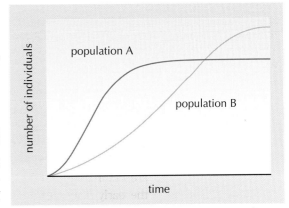

Figure 3.12 Two growth patterns of hypothetical populations.

Comment

Population A has a higher value of r, as the population is increasing more rapidly than population B. Population B has a higher value of K than population A, as it has stabilized at a higher number of individuals (i.e. its carrying capacity is higher).

Maximum sustainable yield

Understanding population growth is very important when we wish to regulate exploitation (hunting or fishing) of wild populations because, while we may know the *relative* rate at which individuals reproduce themselves, we will have no inkling about the population size unless we have some information about the *absolute* rate of increase. The total size of a population is critical to its long-term survival, with small populations generally less stable than large populations (**Freeland, 2003**). The key to sustaining hunted or fished populations is knowing when to stop hunting or fishing. It is obvious that if we continue to remove individuals from a population at a rate that is greater than their replacement the species will eventually go extinct, a pattern that has been repeated on many occasions, with some famous examples including the passenger pigeons (*Ectopistes migratoris*) in North America (**Hinchliffe and Belshaw, 2003**) and the dodos (*Didus ineptus*) on Mauritius. At the same time, many people wish to maximize the yield that they can obtain from populations such as cod (*Gadus* spp.), haddock (*Melanogrammus aeglefinus*) and many other species that provide food or other benefit. So how do we know how many individuals we can remove from a population without driving it to extinction? This balance between protecting the population and harvesting as many individuals as possible is known as the **maximum sustainable yield (MSY)** .

maximum sustainable yield (MSY)

We can understand the concept of MSY by returning to our deer population and factoring in this population's carrying capacity. As populations start to approach their carrying capacity, both absolute and relative growth rates will slow down until growth becomes zero. At this point (at carrying capacity) the deer are still reproducing, but birth plus immigration does not exceed death plus emigration. Referring back once again to Figure 3.11, we are reminded of how a growing population will generally undergo an increase in its absolute growth rate until it approaches its carrying capacity; that is, the growth rate starts to decrease *before* carrying capacity is reached. You can see from Figure 3.11 that the steep slope of stage 2 growth starts to level out before the plateau of stage 3, reflecting a gradual slowing in growth rate. Population growth is generally most rapid at about half of the population's carrying capacity, because at this time there will be enough individuals in the population for rapid absolute growth, and there should be no limitations of food and other resources. Therefore, at least in theory, if a deer population is continually harvested to one half of its carrying capacity then this should provide the maximum yield without compromising the population's ability to perpetuate itself.

Activity 3.6

What reasons can you think of that make harvesting a population to half its carrying capacity an inexact science?

Comment

As we have learned, populations do not remain at constant sizes from one year to the next. For example, food availability fluctuates, and there can also be fairly pronounced size reductions following disease, weather extremes, etc. If there is an outbreak of disease in one year, harvesting that is intended to reduce a population to half of its carrying capacity may actually cause a much bigger reduction from which the population will have a harder time recovering. In addition, carrying capacity will depend to some extent on fluctuations within populations of other species (e.g. prey species, as we saw in the lynx–snowshoe hare example).

Maximum sustainable yield is, in reality, extremely difficult to calculate. In addition to fluctuations in population size and carrying capacity, factors such as the age of first reproduction, proportion of immature versus mature individuals, and longevity (how long individuals live) must be taken into account. For example, the MSY of African elephants is considered to be only about 5 per cent of the population, because reproduction in this species is slow – age of first reproduction is 10–11 years, and the gestation period is 22 months. In other species (such as faster growing ones) the MSY will be higher; the importance of MSY and cod fisheries is discussed in Chapter Five.

○ Why do fishing restrictions often specify not just the limits to numbers that can be caught, but also the size limits below which fish must not be harvested?

● If small (that is, young) individuals are removed from a population then, as the older ones die, there won't be enough young ones to replace them. When this replacement ability is lost, the population will have no way to replenish itself and will be in danger of total collapse.

We will now return once more to what is probably the most prevalent *K*-species on our planet – *Homo sapiens.*

4.2 Human populations

When considering the patterns of human population growth, one is struck by the remarks above about *r*- and *K*-species. You will recall that the growth and reproductive strategy of the former is called an *r* strategy. The individuals in *r*-species 'start to reproduce at an early age and have a large number of offspring ... The cost of this strategy is a high mortality rate – many offspring will die as they are not well provided for'. This sounds very much like the situation of human populations as expounded by the Revd Thomas Malthus over two centuries ago (see Box 3.2 and Figure 3.13). However, in later editions of his essay, Malthus laid much more stress on a reproductive strategy that, as we noted earlier, was said to characterize K-species. Here 'age at first reproduction ... is later than in r-species and ... more investment is placed in each offspring'.

Box 3.2 Malthus and the principle of population

The Revd T.R. Malthus (1766–1834) is remembered for his *Essay on the Principle of Population*, which was first published anonymously in 1797. In it he took issue with those who had an optimistic view of the future of humankind (Jean Jacques Rousseau among them), arguing that the natural tendency of population to increase faster than the means of subsistence would make life for most people one of constant struggle. His work was taken up politically by those who were aghast at the revolution in France, including its philosophical underpinning, and who railed against the rapidly rising cost of handouts to the poor at home. Such handouts tended, not surprisingly, to be linked to the size of families. Thus, a married couple with a large number of children would get more than one with a small number. A single man or woman would get less still. Malthus argues that this encouraged people to marry too early, before they could properly support a family. In 1834 a new Poor Law was passed. The main thrust of this was that relief would now only be given to people who entered a workhouse. Conditions in workhouses were harsh. Families were split and put into different wards for men, women and children. This according to Malthus and his supporters would discourage early marriage.

Figure 3.13 Portrait of Revd T.R. Malthus by J. Linnel, 1823.

Table 3.3   Married women as a percentage of the women in each age group

	15–19	20–24	25–29	30–34	35–39	40–44	45–49
India, 1931	84	90	87	83	70	62	47
Sweden, 1750	4	27	56	71	79	79	75

Source: Krause, 1958–1959, p.172, reproduced in Drake, 1973.

○   In the light of the above remarks, how would you interpret Table 3.3?

●   It is immediately apparent that women in India in 1931 were (up to age 35 at least) far more likely to be living in the married state and therefore exposed to child bearing, than were their counterparts in Sweden in 1750. At the earlier ages (15–19, 20–24, 25–29) the differences were enormous. There was a reversal of the position at the later ages, brought about by higher death rates in India.

Malthus put forward two propositions: 'first, that food is necessary to the existence of man [in Malthus's day, 'man' stood for 'humans']. Secondly, that the passion between the sexes is necessary and will remain nearly in its present state'. He then argued that 'the power of population is indefinitely greater than the power in the earth to produce subsistence for man'. Population when unchecked increases in a geometric ratio [1, 2, 4, 8, 16, 32 and so on]. Subsistence increases only in an arithmetic ratio [1, 2, 3, 4, 5, 6 and so on]. Malthus went on to demonstrate that these 'two unequal powers must be kept equal'. This, he argued, was brought about by two kinds of checks: the **positive check**, which cut back a population 'increase which is already begun, e.g. war, famine, epidemics, vice' (abortion, infanticide), and the **preventive check**, which arose from 'a foresight of the difficulties attending [the rearing of] a family' (Malthus, 1798, pp.11, 13, 14, 90). Figure 3.14 presents these checks in a diagrammatic form.

positive check

preventive check

Figure 3.14 consists of two causal loops: the outer (purple) one represents the operation of the 'positive' check associated with food supply, the inner (blue) one the 'preventive' check associated with marriage. We start with 'population size'. If the population is increasing, then, in a malthusian world, food prices will tend to rise owing to humanity's inability to raise food output as rapidly as it can reproduce itself (the working of the arithmetic rate as against the geometric rates of growth). Thus there is a *positive correlation* between population growth and the rise in food prices; that is, they both go up together. This correlation is represented by the plus sign on the diagram. If food prices rise, then, other things being equal, *real income* will fall. This is a *negative correlation* (i.e. one going up and the other going down) and is represented by a minus sign on the diagram. If the fall in real income is profound and/or prolonged, then food

Positive checks in non-human populations would include famine, drought, disease and climatic extremes.

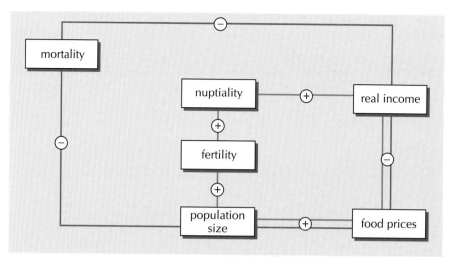

Figure 3.14 The positive and preventive checks.
Source: Wrigley and Schofield, 1989, p.458.

consumption must be curtailed, which will lead to a rise in mortality, either through diseases associated with malnutrition or, in extreme cases, through famine – another negative correlation. A rise in mortality will lead to a reduction in the rate of population growth or, if it is high enough, to an actual fall in population. This, in turn, will lead to a fall in food prices, a rise in real incomes, a fall in mortality and an increase in the rate of population growth and so on and so on.

In Figure 3.14 the inner loop or circuit shares the initial linkages with the outer one, that is, population size to food prices to real income. If real income falls, then, according to Malthus, people will be hesitant about embarking upon marriage. Thus nuptiality – the proportion of the population living in the married state – will fall; the age at marriage is likely to rise, so the amount of time a woman is exposed to child-bearing is also likely to fall, which, in turn, will reduce the fertility of the population. A fall in fertility means a reduction in the number of new members being added to the population, and, therefore, a fall in its rate of growth. Conversely, a rise in real income will encourage more people to get married and at an earlier age, which, in turn, will lead to higher fertility and, other things being equal, an increase in population size.

Malthus argued that humans produced more children when 'food' was plentiful, at the time of rising real incomes. In subsequent editions of the *Essay*, Malthus drew on a great deal of empirical evidence, both historical and contemporary, to support this argument. Malthus was speaking of a pre-industrial world and so far as pre-industrial England is concerned he seems to have been right (Wrigley and Schofield, 1989). One should also note that although men who were better off at that time tended to marry later than those less well off (e.g. farmers as against farm labourers) they tended to choose younger brides. The result of this was that the former had larger families. More recently, in the post-industrial world, children have become a 'consumption good', analogous to cars, houses, etc., rather than an 'investment good' (an extra hand for the farm or workshop and a support in old age). More recently still, however, the increased opportunities for worthwhile careers for educated women has led to delays in starting families and, in an increasing number of cases, to a decision not to have children at all.

Malthus has been in the firing line ever since he put forward his theory. Those he sought to refute did not take kindly to his writings. They believed in the perfectibility of human beings and in progress to a better world. Malthus believed there was no such thing. The two checks were ever present. Both involved pain and sacrifice. After his death, advances in technology, principally in the field of transport, resulted in a situation where, in western Europe at least, rising populations did not trigger rising food prices, as they had done previously. The steamship and the steam railway led to the opening of the American prairies, the Argentine pampas, Australia and New Zealand to the pastoralist and agriculturalist. This meant that western populations could grow and be fed. Malthus did foresee some grounds for optimism. In fact he believed that populations could double every 25 years if they enjoyed optimal conditions of abundant land and a social system that allowed it to be exploited. He identified the newly founded USA as providing such an optimal environment.

Cheap and reliable mechanical forms of birth control came in the twentieth century. If they did not douse the passion between the sexes, they certainly curtailed its effects. It should be noted, however, that in England such devices only came into widespread use after the Second World War. The major reduction in fertility that occurred between approximately 1870 and 1930 was brought about by abstinence from sexual intercourse or *coitus interruptus*. It is, however, worth noting that abortion (which Malthus very definitely labelled a vice and many people still would) remains a major force in the control of population. Currently, in the UK one in three women are expected to have at least one abortion before reaching the age of 45 years, with an estimated one in five conceptions deliberately aborted.

What is the bearing of Malthus's theory on our concerns? Put simply, it would appear that the history of the human species up to the time of his writing had shown that population growth was slow over the long term because technological advances were not sufficient to raise food supply to meet the needs of a growing population. Such needs could only be met by taking in more land. This was a laborious process, hence the shift from hunter-gathering to pastoralism to settled agriculture was slow and as a result so was population growth. From the year 1000 onwards, we find that over the short to medium term, populations might grow relatively rapidly, only to be cut back by the effects of malnutrition, or in the worst case, by starvation, in a manner akin to that shown above in Figure 3.8 for non-human populations. Such changes were often related to favourable or unfavourable weather conditions. Alternatively population growth could be held back by preventive checks (late age at marriage or permanent celibacy) and would come to a halt when the carrying capacity of the area occupied by the population was reached. In either case the impact on the environment would be one of slow change.

Malthus was not unaware of technological change in agriculture. For instance, in the second paragraph of his *A Summary View of the Principle of Population,* he noted that when wheat 'is dibbled instead of being sown in the common way [broadcast] two pecks of seed wheat will yield as large a crop as two bushels, and thus quadruple the proportion of the return to the quantity of seed put into the ground' (Malthus, 1830, p.119). But, as the quotation shows, the drilling (dibbling) of wheat was uncommon and the technique of broadcasting seed, with all its shortcomings so graphically described in the Bible, was the norm in what was, in 1830 (the year *A Summary View* was published), the most advanced agricultural economy in the world.

○   Can you think of a major fall in Europe's population in the period 1000–1500 AD?

●   There was a major fall in population in the period of the Black Death caused by bubonic plague, principally in the years 1348–1352 (**Blowers and Smith, 2003**) but also throughout the fifteenth century. There is no apparent link between food supply and bubonic plague as such. Other diseases had direct or indirect links with foodstuffs. In addition, famines on a wide scale occurred in the run up to the Black Death from 1300 to 1350.

Summary

In Section 4.1 we have learned that variation in population sizes is generally most pronounced in *r*-species, because their ability to reproduce rapidly means that they are often well adapted to temporary environments. In contrast, populations of *K*-species tend to remain relatively stable over time but, as a result of this, *K*-species take longer to respond to environmental perturbations (including over-hunting and over-fishing) that result in decreased population sizes. For this reason, estimates of MSY are often desirable, although changes in a multitude of variables mean that accurate MSY predictions are extremely difficult, if not impossible, to maintain from one year to the next.

The human population theory propounded by Malthus has been highlighted here, in part because, according to him, human population numbers were held back in two ways. These bear striking similarities to the checks operating on non-human populations. The first he called the 'positive' check which cut back a population that had grown beyond its long-term capacity to maintain itself. The second he called the 'preventive' check, which operated to restrict population growth so as to maintain living standards.

They are of interest for two reasons. First they help to explain why human populations have grown so slowly *until relatively recent times*, which, by extension, has meant that their impact on environmental change was relatively modest. Second, the dynamics of human populations that underlie the 'positive' and 'preventive' checks, have important implications for economic growth since the Industrial Revolution. As Section 6 will show, this economic growth has been one of the most important – if not the most important – determinant of environmental change in our times.

5   Human versus non-human populations: conflicts and change

It will be apparent from this and earlier chapters, that since the arrival of *Homo sapiens,* the physical environment has been a battleground for humans and non-humans. Among hunter-gatherers, the centrality of the contest takes a graphic form in the dominant motif in cave drawings that have been found in various parts of the world (Figure 3.15). The extension of that conflict across the globe can be deduced from the migratory paths of human populations (Figure 3.16).

Figure 3.15 Cave drawing showing hunting of animals: the conflict between humans and non-humans has a very long history.
Source: Wells, 1925, p.53.

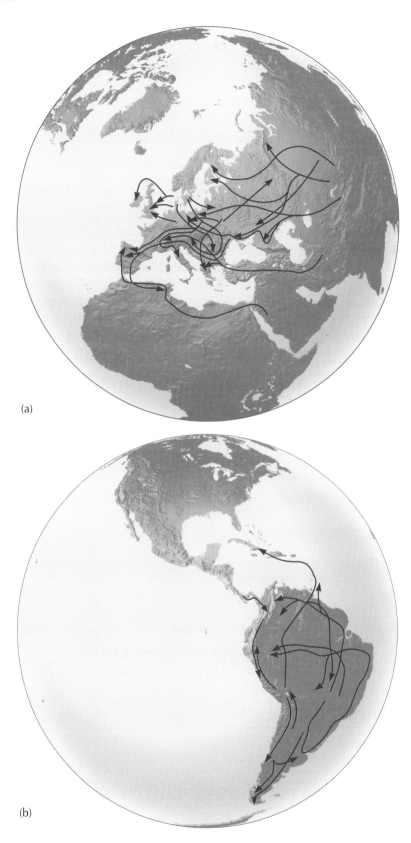

(a)

(b)

(c)

Figure 3.16 Migratory paths of humans across globe from earliest times: (a) across Europe, (b) across South America and (c) across Asia.
Source: Debenham, 1960, pp.42, 86, 118.

The occurrence of such migrations suggests that, over time, human populations prevailed, although to the individual hunter, at any one time, this probably didn't appear likely. The partial shift of the world's human populations from hunter-gathering to slash-and-burn and then to settled agriculture implies that environments were, on the whole, increasingly won over by human populations at the expense of non-human ones.

○   Can you think of any human actions that led to the preservation
       of non-human populations in the process just described?

●   There may be more than one answer to this question, but the most
       obvious one that we can think of is the domestication of animals.

The contests between human and non-human populations that occurred in pre-historical times are poorly understood as we have few details of them. From the growth of human populations, we can assume that the pace has quickened since 1650, and even more so since 1950. There has, however, been an interesting change in the nature of the contest since 1950. It is reckoned that between 1700 and 1960, the farmland of the world increased fourfold (Richards, 1990, p.164; FAO, 2000). This was just about the same rate of growth as that of the world's population (see Figure 3.5). This might suggest that over that time the supply of food had been increased in the traditional way – by taking in new land for cultivation from forests, marshes and grassland. This was probably not the whole story, since there had been technological changes (e.g. fertilizers, irrigation and improved strains of seeds and animals) which raised agricultural productivity and so added to the supply of food. These two developments in tandem probably led to many in the world being better fed in 1950 than in 1700, both in terms of quantity and quality.

What has happened since then? The world's population has more than doubled, yet the farmland area has increased by only 12 per cent (Goklany, 1998; IFA, 2000). See Box 3.3.

Box 3.3 Population growth and land use

A 400 per cent increase in farmland between 1700 and 1961, and a 12 per cent increase between 1961 and 1998 might suggest a slowing down in the rate of increase. But is there more to it than that?

In 1700 the area of the world's farmland has been estimated at 265 Mha. By 1961 it is estimated to have been 1,364 Mha – a rise of 400 per cent. By 1998 the farmland area was estimated at 1,512 Mha – an increase since 1961 of 12 per cent. There were 261 years between 1700 and 1961, so in that time the area of farmland grew by an average of 4.14 Mha a year. There were 27 years between 1961 and 1998 and in that time the average increase in farmland was 4.5 Mha a year – so not much difference there. However, as we have seen, the absolute increase in population per year was much greater between 1961 and 1998 than it had been between 1700 and 1961. This means that although the amount of new land taken into cultivation was about the same in the two periods, the amount *per head* was much less. The age-old link between population growth and the taking in of new land – as we saw earlier, the core relationship according to Malthus – had been broken.

Over the period 1961 to 1998, not only had most of the world's population been fed, the energy content of that food intake had increased. In the world as a whole, the daily intake of food energy per capita has grown by 24 per cent and in the developing world by 38 per cent. The increase in the developed world has been a more modest 12 per cent (FAO, 2000).

Activity 3.7

Why has the rate of growth of daily food energy intake been so much lower in the developed world than elsewhere and what environmental impacts have resulted from what growth there has been?

Comment

Populations in the developed world were already comparatively well fed in 1961. At the present time, if anything, they are over-fed, with obesity becoming a major health problem. The growth in consumption has been, in part, a case of 'more of the same'. More significant, however, has been the growth in the consumption of convenience foods. The development of such foods, together with the globalization of food supplies to the developed world (an enormous speeding up of the process that experienced take-off in the mid-nineteenth century) has had an environmental impact. This has resulted from the much longer distances food now travels.

To help you think further about the globalization and increased production of food supplies, turn now to Activity 3.8.

Activity 3.8

Next time you empty your food shopping bags, try to estimate how far each of the items has travelled from where it was manufactured/grown to your home. As a benchmark, you might compare your findings with the suggestion that the average item on the supermarket shelf has travelled 1,000 miles between producer and consumer.

How has the increase in world food supply been brought about? In part the increase has come the traditional way, from more land being taken into cultivation. But much has resulted from better seeds and animals, irrigation, pesticides and artificial fertilizers. We can take the last of these, by way of illustration. Although the scientific breakthrough for artificial fertilizers came in 1908 (from the work of Fritz Haber), and it was feasible to produce the fertilizers on an industrial scale shortly afterwards in 1914 (as a result of the work of Karl Borch), they were not widely adopted for another 50 years. In the world as a whole the use of artificial fertilizers rose from 5 kg per head in 1950 to 28 kg per head in 1990. It then fell, principally as a result of the collapse of Russian agriculture, only to resume its upward trend from the mid-1990s. In the developing world there has been no decline and, not surprisingly, the rate of growth has been faster, from 2 kg per capita in 1962 to 18 kg in 1999 (IFA, 2000; Brown et al., 1999)

A final question, Activity 3.9, should help you to understand how the developments in human food supplies have affected non-human populations.

Activity 3.9

How have the technological changes since about 1950 impacted on the contest between human and non-human populations?

Comment

We have seen that non-human populations have been under considerable pressure for territory, with the amount of new land being taken into cultivation per year remaining, on average, at the same level as during the previous 250 years. Given the vastly increased rate of growth of human populations, the situation could have been much worse! Non-human populations have, however, had to contend with a wholly new threat, namely the massive increase in the use of herbicides and pesticides. Examples of increased yields brought about by the use of pesticides, amongst other things, are legion. For example, one-third of the world's rice output is said to have been eaten by insects in 1960 (Heinrichs, 1998). Closer to home, anyone who can remember the English countryside in the 1940s and 1950s will remember the presence of enormous numbers of rabbits and the damage they caused to crops for many metres of ground alongside hedgerows. The introduction of myxomatosis, with its horrible symptoms, transformed the face of the countryside (Figure 3.17) as well as increasing output.

There has, however, been a downside for both human and non-human populations in the widespread use of artificial fertilizers, herbicides and pesticides. That lies in their side-effects, principally water pollution caused by oxygen depletion and the increase in nitrate and other substances (see Chapter Four). Box 3.4 gives examples of some specific effects of the insecticide DDT (dichlorodiphenyltrichloroethane).

Figure 3.17 Rabbit damage to a barley field in Kent: (a) 1954, before myxomatosis, one third of the crop lost, and (b) a good crop of wheat in the same field in 1955 after myxamatosis.
Source: Thompson, 1994, p.91.

Box 3.4 DDT (dichlorodiphenyltrichloroethane)

In the 1960s the insecticide DDT was implicated in decreasing population sizes of many birds of prey including ospreys (*Pandion haliaetus*, Figure 3.18), bald eagles (*Haliaetus lencocephalus*) and peregrine falcons (*Falco peregrinus*). These birds are high in the food chain (Chapter One). Each link in the food chain leads to an increased concentration of environmental substances because, for example, a falcon will be consuming the DDT that has been acquired by a shrew (falcon's prey) plus that acquired by many insects (shrew's prey) and other invertebrates and plants (insects' food). This increased concentration through the food chain is called **bioaccumulation**. The high levels of DDT in birds of prey made their egg shells weak and thin, and as a result eggs often collapsed under the weight of parents. The subsequent reduction in birth rates was substantial, resulting in many instances of local extinction. More recently, it has been discovered that DDT is one of several chemicals that can mimic oestrogen, a hormone that is found predominantly in females. These so-called environmental oestrogens have been implicated in a variety of phenomena including the feminization of alligators (so that some populations contained relatively few functional males), and the reduction in sperm counts in humans. DDT has been banned from agricultural use in most countries since the 1970s, but can last a very long time in the environment; half of the DDT in soil will take 2–15 years to break down. Furthermore, it is a cheap insecticide and in some developing countries it continues to be used in limited quantities for public health purposes, particularly for controlling mosquito populations and malaria.

bioaccumulation

Figure 3.18 Ospreys (*Pandion haliaetus*) on nest, Florida: ospreys have been victims of DDT applications, with egg shells unable to withstand the weight of the parent on the nest.

Summary

One wonders what Malthus would have made of it! The age-old contests continue: that between human and non-human populations and that between human populations and the environment from which they attempt to wrest a living. But the new threats to all populations are of a different order from those Malthus envisaged and all their outcomes are, as yet, unknown. Examples would include the impacts of air and water pollution, the effects of pesticide residues and so on – all of which can have deleterious effects on human and non-human health.

6 Population growth and economic growth

In what follows we are using economic growth as a surrogate for environmental change. We can do this because in the modern context, the environmental changes associated with economic growth have been both massive and multi-faceted. As long as raw materials and sources of heat and power were largely organic in origin – wool, hides, wood, animals and the work done by humans – there were limits to growth, and expansion in output (by post-Industrial Revolution standards) was slow and hard won. Environmental change was similarly slow and hard won. Economic growth had been constrained by the fixed supply of land – the ultimate source not only of food, but of raw materials and energy. The move to mineral sources of raw materials and the equally important use of minerals, initially coal, for energy, pushed back the limits to growth and hence to environmental change (Chapter Two).

How then did population growth relate to economic growth? Did it stimulate or retard it? We'll begin by putting forward three arguments, each of which suggests that rising populations stimulate economic growth.

- The first is that if population growth is translated into effective demand, this encourages investment. By effective demand for an economic good we refer not to the amount that people would like to be able to buy, but to the amount actually purchased at a given price.
- The second is that if rising population leads to a fall in the cost of labour, this will increase profits and so provide investment funds.
- A third argument is that a rising population will lead to an increasing density of population. This opens the way to greater specialization in employment (one person doing one thing rather than many), leading to improvements in efficiency. This, in turn, leads to a rise in productivity (a rise in output per unit of effort) in both agriculture and industry. Also an increasing density of population allows economies of scale, particularly in transport, manufacturing and various service industries.

Each of these operates under the law of increasing returns: the bigger they are (and their size is a function of population size suitably modified by the level of its effective demand), the more efficient they are likely to be.

Against these arguments can be set the following counter-arguments.

- First, a rising population in an underdeveloped economy may not lead to much of an increase in effective demand. By definition income per head will be low in an underdeveloped economy. A rising population will diminish it still further and so reduce whatever savings the mass of the population are capable of making, thus decreasing the supply of investment funds.
- Second, although a rising population in an underdeveloped economy may increase the supply of labour and thereby cut its cost, the amount in either case is likely to be small since there is possibly already a large pool of unemployed or under-employed labour. Furthermore, any rise in incomes is not likely to be transformed into investment funds. This is because in an underdeveloped economy the main recipients are either landowners, whose lifestyle leads them to spend it on immediate and conspicuous consumption, or small traders who get it through their money-lending and short-term credit operations and who are more than likely to use it to purchase land.
- A third counter-argument is that a rising population in an underdeveloped economy is more likely to see the operation of the law of diminishing rather than increasing returns. This is because land, being the main resource in such an underdeveloped economy, will be subdivided; over-cropping will ensue, soil will become exhausted more quickly and the diminishing size of holdings will militate against the use of machinery. Population increases may in these circumstances increase poverty and as **Hinchliffe and Belshaw (2003)** argued, poverty can produce environmental degradation and hence more poverty.

Several other arguments have also been brought forward to suggest that a rising population hinders the economic growth of an underdeveloped economy. First, such an economy will have a **high burden of dependency**, that is to say the high burden of dependency number of children will be large relative to the working population that has to support them. A rise in population will increase this burden, since it will lead to a rise in the number of children. This will mean that even more labour (often particularly female labour) will be tied up in caring for them. Furthermore, resources that might have been used for investment purposes will be needed to satisfy current consumption. Linked with this, funds which might have been used to raise the amount of capital per worker will instead be needed to duplicate existing facilities, whether these are productive capital (i.e. another plough rather than a better plough) or social overhead capital (i.e. another house rather than a better house). It has been demonstrated that the faster a population grows, the more investment is needed to keep constant the amount of capital per worker. In numerical terms, it is reckoned that if a population is increasing by one per cent per annum, an investment of 2.5 to 5 per cent of national income is needed, merely to keep constant the amount of equipment per worker. If the population is growing by, say, 2.5 per cent per annum, then the amount of national income needed to maintain the status quo rises to between 6.25 and 12.5 per cent of national income – a sizeable amount.

Whether a rising population stimulates or inhibits economic growth is, then, a question without an obvious answer. One possible way of resolving the paradox may lie in the mechanics of population growth, since this encapsulates many of the factors that make one society more likely to experience economic growth than another. We are thinking, for example, of such factors as: the attitude towards the future; the concept of achievement; the willingness to forgo present for future gratification; and a fatalistic or a rational approach to life. To show this let us consider two hypothetical countries with alternative mechanisms of population growth.

crude birth rate

crude death rate

Country A is an underdeveloped agrarian economy. It has a **crude birth rate** of 45 per 1000 (i.e. 45 children are born annually for every 1000 of the population) and a **crude death rate** of 40 per 1000. Marriage is universal for women at an early age (mean age at first marriage is 18 years for women) and there is no deliberate control of births within marriage. The high level of mortality is largely produced by endemic or epidemic disease, sometimes assisted by floods, famine and war. Some external agency enters the country and brings about a sharp fall in mortality. Within 20 years the crude death rate has fallen from 40 to 20 per 1000. The agency might be a better transport system, which alleviates local food shortages, or it may be some new medical discovery, which eradicates a particularly destructive disease, for example, smallpox or malaria. However, there is no change in the economy.

Country B is also an underdeveloped agrarian economy. It has a crude birth rate of 30 per 1000 and a crude death rate of 25 per 1000. Although 80 per cent of the female population eventually marry, their mean age at marriage is about 25 years. The women in this economy spend, therefore, on average not much more than half their fertile period actually exposed to child bearing; a period consciously cut down by the practice of birth control within marriage. Marriage is comparatively late and fertility restricted in order to maintain living standards. Suppose that into such a society there comes some technological change, or some greater demand for its products, which together, perhaps, lead to an increased demand for labour, which makes itself felt through higher incomes. This permits people to marry earlier and to support larger families. Over a period of 20 years, the crude birth rate rises to 40 per 1000, the death rate remaining at 25 per 1000.

At the end of this 20-year period, country A is experiencing a population growth of 45 births–20 deaths, i.e. 25 per 1000 per annum (2.5 per cent per annum), and country B one of 40 births–25 deaths, i.e. 15 per 1000 per annum (1.5 per cent per annum). But whereas country A has a growing population because its death rate has been cut by some exogenous factor, country B's population is rising because a growing economy is allowing the population to grow and living standards, measured in terms of per capita income, to rise or at least be maintained. Country A had a higher burden of dependency principally because its birth rate was higher (see above for discussion of the interlinking of population structures and processes) than country B and a lower income per head. There is nothing in the mechanism bringing about its growth of population that will stimulate growth. Country B, however, starts its economic growth from a stronger economic base: a lower burden of dependency, a higher income per head and a population already limiting its growth in order to maintain its living standards.

The relationship between population growth/population dynamics and economic growth is important to us because of its impact on environmental change. Modern economic growth (introduced by the Industrial Revolution) emerged first in west European societies, with Britain as the prime mover. The demographic regime in Britain seems to have been akin to that of country B, as described above. That is to say its population was subject more to preventive than positive checks from at least the seventeenth century onwards. Changes in nuptiality and fertility were, then, more important than changes in mortality in determining the rate of population growth. Such a demographic regime, it has been argued, was more conducive to economic growth than that in non-western societies akin to country A. It is not surprising, therefore, that large-scale environmental changes associated with industrialization and urbanization should occur first in western societies.

Population growth/population dynamics was only one of the factors bringing about the Industrial Revolution and all that has flowed from it. They were, arguably, necessary but not sufficient conditions. The marriage pattern in Britain – relatively late marriage responsive to economic circumstances – was common throughout western Europe in the eighteenth and nineteenth centuries. But the Industrial Revolution occurred first in Britain and not, say, in Sweden (see above for its nuptiality regime in 1750). Similarly, heavy urbanization, which was very much a part of the economic changes wrought by the Industrial Revolution in the nineteenth and twentieth centuries, is now a dominant feature of less developed countries (see Table 3.1 above). In both cases the urbanization has been associated with countries undergoing rapid population growth. In the West it was, for the most part, congruent with economic growth, but in a number of the countries represented in Table 3.1, this has been less so. Thus, as with our opening case study of the open fields/enclosures, the role of population change is not always predictable.

Summary

Economic growth can lead to environmental change, and population growth can affect economic growth through changes in demand, in the efficiency of economic activities and in the cost of labour. It may either encourage or inhibit economic growth, depending on the circumstances involved and what it is that is controlling population growth. Populations that control their growth through preventive checks on fertility will tend to have a lower burden of dependency (a smaller proportion of the population are children) and stronger economic growth, than do populations where positive checks are dominant. Thus the mechanisms involved, as much as population growth itself, have a close relationship to economic growth and concomitant environmental changes. This is illustrated by the relationship between populations and economic growth during the Industrial Revolution in the UK, which differed from that in other contemporary European countries and in some currently developing countries.

7   Conclusion

In this chapter we have tried to show how the part played by populations – both human and non-human – can help us answer our three key questions.

How and why do environments change?

Environments change in part because human and non-human populations grow, recede and exist in a variety of sizes and densities. In order to understand environmental change it is therefore important to understand the size and distribution of populations as well as their dynamics – birth, death and migration rates. The mechanisms linking population change and environmental change are, to some degree, the same for non-humans and humans. One difference is that humans are marked by their greater adaptability to their environments than are non-humans, through the use of an ever-increasing number of technologies. This adaptability has produced a set of environmental changes that impact heavily on both other populations and on aspects of environments. In the past some of these impacts would have checked human population growth either before it took place (Malthus's preventive checks) or after it had occurred (Malthus's positive checks). How such checks will operate in the future is an open question.

How can we represent, model and predict population changes and their consequences for environmental change?

There are various ways of measuring populations and of presenting both their structures and change over time (rates of fertility and mortality, population pyramids, predator–prey models, r and K strategies of reproduction). The consequences of human population change on environments are difficult to predict as they are, in part, determined by the economy, culture and technology of the particular population.

How do we determine the consequences and significance of population change on environments?

In terms of populations themselves, changes become particularly significant when populations reach the point at which deaths and emigration exceed births and immigration or when they are harvested/preyed upon to such an extent that they cannot replace themselves.

In terms of environments, population changes become significant when carrying capacities are exceeded or when a population decline starts to affect other components of the environment. The significance of population changes depends in part on whether the species or, in the case of humans, the society, follows an *r* or *K* strategy in terms of its reproduction. Partly because of their greater adaptability and the extension of economic systems to other parts of the world, the kind of environmental change that accompanies human population shifts cannot be read off in a simple fashion. Population increases or decreases can contribute to the same type of environmental change (see our example of the field systems of the English Midlands). Furthermore, if population changes are studied in relation to economic growth, then an increasing population in one context will stimulate economic growth, while in another context it will impede it.

References

Anderson, M. (ed.) (1996) *British Population History from the Black Death to the Present Day*, Cambridge, Cambridge University Press.

Blowers, A.T. and Smith, S.G. (2003) 'Introducing environmental issues: the environment of an estuary' in Hinchliffe et al. (eds).

Brown, L.R. et al. (1999) *Vital Signs 1999*, New York, W.W. Norton.

Debenham, F. (1960) *Discovery and Exploration: An Atlas History of Man's Journey into the Unknown*, London, Paul Hamlyn.

FAO (2000) http://apps.fao.ag/

Freeland, J.R. (2003) 'Are too many species going extinct? Environmental change in time and space' in Hinchliffe et al. (eds).

Goklany, I.M. (1998) 'Saving habitat and conserving bio-diversity on a crowded planet' *BioScience*, vol.48, no.11, pp.941–53.

Hall, D., Harding, R. and Putt, C. (1988) *Raunds: Picturing the Past*, F.W. March and Company and Buscott Publications.

Heinrichs, E.A. (1998) *Management of Rice Insect Pests*, University of Nebraska.

Hinchliffe, S.J. and Belshaw, C.D. (eds) (2003) 'Who cares? Values, power and action' in Hinchliffe et al. (eds).

Hinchliffe, S.J., Blowers, A.T., and Freeland, J.R. (eds) (2003) *Understanding Environmental Issues*, Chichester, John Wiley & Sons/The Open University (Book 1 in this series).

IFA (2000) Fertilizer statistical database International Fertilizer Industry Association, http://www.fertilizer.org/stats.htm

Johns Hopkins Bloomberg School of Public Health (2000) *Population Report*, Series M, Number 12.

Krause, J.T. (1958–1959) 'Some implications of recent work in historical demography', *Comparative Studies I: Society and History*, I; reprinted in Drake, M. (ed) (1973) *Applied Historical Studies: An Introductory Reader*, London, Methuen & Co. Ltd in association with The Open University.

MacHutchon, A.G. (1996) *Grizzly Bear Habitat Use Study*, Ivvavik National Park, Inuvik, Yukon, Parks Canada.

MacLulich, D.A. (1937) *Fluctuation in the Numbers of the Varying Hare (*Lepus americanus*)*, Toronto, University of Toronto Press.

Malthus, T.R. (1798) *An Essay on the Principle of Population*; reprinted (1926) *Essay Concerning Human Populations, 1798*, London, Macmillan & Co. Ltd.

Malthus, T.R. (1830) *A Summary View of the Principle of Population*, London, John Murray; reprinted in Wrigley, E.A. and Souden, D. (eds) (1986) *The Works of Thomas Robert Malthus*, London, Pickering..

Overton, M. (1996) *Agricultural Revolution in England. The Transformation of the Agrarian Economy 1500–1850*, Cambridge, Cambridge University Press.

Population Information Program (2000) *Population and the Environment: The Global Challenge*, Series M, No.15, Center for Communication Programs, The Johns Hopkins Bloomberg School of Public Health. www.jyuccp.org/pr/m15edsum.stm (accessed July 2002).

Richards, J.F. (1990) 'Land transformation' in Turner, B.L. (ed.) *The Earth Transformed by Human Action*, Cambridge, Cambridge University Press.

Slack, P. (2002) 'Comment: perceptions and people' in Slack, P. and Ward, R. (eds) *The Peopling of Britain: The Shaping of a Human Landscape*, The Linacre Lectures 1999, Oxford, Oxford University Press.

Thompson, H.V. (1994) 'The rabbit in Britain' in Thompson, H.V. and King, C.M. (eds) *The European Rabbit: The History of a Successful Colonizer*, Oxford, Oxford University Press.

UN (2000) *Population Division Report*, New York, UN, March.

Wells, H.G. (1925) *The Outline of History*, London, Cassell & Co. Ltd.

Wilkinson, R.G. (1973) *Poverty and Progress: An Ecological Model of Economic Development*, London, Methuen.

Wrigley, E.A. and Schofield, R.S. (1989) *The Population of England 1541–1871: A Reconstruction*, Cambridge, Cambridge University Press.

Chapter Four

Changing land

Dick Morris

Contents

1 Introduction

The preceding chapters have examined various processes of environmental change, as well as some models of the overall impacts that human activities have on environments. This chapter examines human impacts on one particular part of our whole environment – **land**. For present purposes, land is defined as the area of the Earth's surface that is not covered by water, but can include smaller lakes and rivers.

land

○ Have a look around you. How much of your lifestyle is dependent on products (in the widest sense) that are derived from land?

● Obviously your answer will vary with your particular circumstances, but we are all dependent on food, most of which is grown or reared on the land. The table under my keyboard is made of wood, another land-derived product, and the building in which I am sitting occupies land, as do the roads we travel along to get to work, shops and so on. My computer keyboard is plastic, made from fossil fuel, with some metal: metals and some fossil fuels are extracted from underneath the land. I also use land for recreation, walking on fields and moorland when possible.

In truth, our very lives depend on land, so the way in which we use and change it has to be important. This chapter aims to show how and why the nature of land varies as a result of human activities and other processes and to examine one way of assessing human environmental impact using land area as an indicator. In particular, it will consider the impact of agricultural activities, since these represent the largest single use of land by humans, but it also looks at the extraction of mineral resources, which can have major localized and sometimes wider-scale effects on the land. The chapter looks at the way in which land varies across the world and over time, and considers some of the uncertainties involved in statements about these changes. It also considers some aspects of the unevenness of human access to land or to the resources that can be extracted from the land.

The key questions that underpin this chapter are:

• How and why does land change, and why is this important?
• How can we measure changes that affect the nature of land?

2 Land and life

At the most basic level, as Chapters One and Two showed, almost the whole functioning of the biosphere depends on conversion of solar energy into stored chemical energy that can be used by living organisms. So, for our global ecosystem to function, the primary producers that undertake this conversion require a surface area exposed to sunlight on which to display their leaves or other photosynthesizing tissue. The surface could be land or sea, although the potential primary production from a given area of open ocean is usually less than on a similar area of land. Some of the light hitting the ocean is absorbed by the water, leaving less for photosynthesis, and the supply of nutrients is usually limited. Interestingly, at the junction between land and sea, estuaries like the Blackwater (**Blowers and Smith, 2003**) have some of the highest rates of primary production of areas that are not managed specifically by humans.

○   From what you learned in preceding chapters, why might estuaries have high rates of primary production?

●   The main reason is that the resources needed for growth (water, light and plant nutrients) are readily available. The river supplies fresh water and nutrients in the form of soluble or decomposable materials. These are also brought in with the tide from the sea. So photosynthesis can potentially take place at its maximum (primarily light-limited) rate.

However, land and sea are much more than just solar collectors, and this is reflected in the economic value that we place on land as **property**. As defined by economists, a **property right** enables the owners of a particular resource to use it in whatever way they choose. The idea that land can be owned by individuals, as private property, is deeply ingrained in western thought, to the extent that the phrase 'property sales' is in the UK synonymous with sales of land itself, or a dwelling on it. However, many of the aboriginal peoples inhabiting areas prior to their 'discovery' by Europeans had very different relationships to land. Even within Europe there are still common lands (mentioned in the previous chapter), where a group, rather than any individual, has rights over that land. Such common rights are often restricted. Where the vegetation on the land is regarded as common property, commoners may only have the right to graze a limited number of livestock. More generally, even private property rights relating to land are not necessarily absolute.

property

property right

○   From your experience, suggest some ways in which the rights of individuals with regard to land they own may be limited.

●   Planning regulations in many states limit the freedom of property-owners to use their land for certain activities such as building on it. On enclosed agricultural land, there may be rights of access for walking, so the owner cannot legally exclude walkers from the property. The property rights to mineral resources under the land may also be separate from those associated with surface features.

Where land is private property, it has a financial value. Table 4.1 shows the value in pounds sterling of land in various different areas of the UK in 2001. Values are given for four different possible uses: for building of dwellings; for industrial workplaces, roads and other infrastructure; for crop production; and for extraction of materials from below the land.

Activity 4.1

Look at Table 4.1. What are the striking features of the information contained in this table?

Consider your answer before reading the text that follows.

Table 4.1   Value of land in pounds sterling per hectare[1] in different areas of the UK, for different possible uses, October 2001

	Residential land	Industrial land	Arable agriculture	Sand and gravel extraction
London Inner				
Camden	12,500,000	–	–	–
Southwark	3,500,000	750,000	–	–
South East	–	–	2,780	40,000–130,000
Eastbourne	1,050,000	375,000	–	–
Oxford	7,500,000	950,000	–	–
North East	–	–	7,842	20,000–28,000
Middlesbrough	450,000	130,000	–	–
North West	–	–	8,028	40,000–50,000
Bolton	600,000	275,000	–	–
Manchester	1,100,000	400,000	–	–
East Midlands	–	–	7,242	30,000–47,000
Lincoln	575,000	250,000	–	–
Loughborough	1,450,000	555,000	–	–
South West	–	–	7,677	20,000–25,000
Exeter	1,500,000	525,000	–	–
Plymouth	725,000	295,000	–	–
East Anglia	–	–	7,711	30,000–50,000
Cambridge	2,275,000	650,000	–	–
Colchester	1,250,000	600,000	–	–
West Midlands	–	–	8,316	35,000–50,000
Birmingham	1,500,000	700,000	–	–
Stoke-on-Trent	450,000	250,000	–	–
Wales	–	–	–	20,000–25,000
Cardiff	1,850,000	370,000	–	–
Carmarthen	225,000	150,000	–	–
Scotland			5,664	17,500–35,000
Edinburgh	3,000,000	275,000	–	–
Dumfries	176,000	110,000	–	–
Northern Ireland	–	–	9,922	–

Note: [1]One hectare is an area 100 m x 100 m.

Source: The Property Market Report, Autumn 2001. Valuation Office Agency, Crown Copyright; *Estates Gazette*.

Table 4.1 demonstrates that the price of land across the UK varies widely with location and with the use to which it can be put. Residential land can be one hundred or more times more valuable than arable land in any region, and its price can vary by a factor of ten between regions. The reduction in value of residential land with distance from London or other centres of commercial activity is also particularly extreme. The UK is unusual, in that more housing is owned by its occupier than is the norm in the rest of Europe. Land values in the UK are among the highest in the world (even one hectare of arable land can be worth half the average annual wage), but almost everywhere land is regarded as a valuable commodity.

○   Does the financial value of land shown in Table 4.1 relate in any way to its importance as a site for photosynthesis?

●   Apparently, not at all. Land which is primarily for built structures or for gravel extraction is many times more valuable than arable land.

Freeland (2003) and Chapter One showed that biodiversity is affected by the nature and area of land available to non-human organisms. This is also not usually reflected in any way in land prices, nor are other aspects of aesthetic 'value' **(Hinchliffe and Belshaw, 2003)**. Since the late nineteenth century the mountainous landforms of areas such as the Alps and the English Lake District have been greatly appreciated by people who do not have any formal property rights to them – for their scenic beauty and for the opportunities for recreation that they offer. In many cases, there is potential conflict between

Figure 4.1 Limestone quarrying in the English Peak District.

use of land as an economic resource for mineral extraction and its aesthetic value: consider the example of the Derbyshire Peak District shown in Figure 4.1. The rest of this chapter looks particularly at humans' economically important activities on land, and the ways in which these can affect other aspects of environments.

Summary

Land is essential to human survival, to provide food, to supply places on which to build and as a source of some other resources. Where land is held as private property, its value does not seem to relate in any way to its biological role. This may affect the way in which we change land as part of our environments.

3   Land on a global scale

Table 4.1 represented one way of looking at land, according to its value, based on its use and spatial location. Another way of categorizing land is in terms of the different biomes (**Freeland, 2003,** and Chapter One) whose features are determined by their climate, geology and history. The biomes reflect processes independent of human action, but human activities now have a major effect on

Figure 4.2 Some typical arable crops: (a) wheat (*Triticum aestivum*); (b) sugar beet (*Beta vulgaris*); (c) maize (*Zea mays*).

land and Table 4.2 categorizes global land area to reflect this. Much land is used for a variety of agricultural activities, including arable crops such as grains, soya or specially sown grasses (see Figure 4.2). A relatively small area is used for permanent woody crops such as coffee, tea, cotton and some nuts (see Figure 4.3) and a much larger area is managed grassland. This last area, so-called permanent pasture, may look much like the steppe or savannah biome, but contains different species and may occur in climatic regions where these biomes would not naturally be found. In all grasslands, the processes of succession (Chapter One) are checked by climate, fire, grazing or human activity to prevent the grassland being replaced by more woody vegetation.

Table 4.2 Global percentages of land in different categories, 1991

Arable land and permanent crops	11.18
Permanent pasture	25.40
Forest and woodland	32.15
Other land	31.27
Total area of land in use (ha x 10^3)	13,426,030

Source: Data from FAOSTAT.

(a)

(b)   (c)

Figure 4.3 Some examples of permanent crops: (a) tea (*Camellia sinensis*); (b) coffee (*Coffea arabica*); (c) cotton (*Gossypium hirsutum*).

Figure 4.4 Examples of plantation forestry: (left) in the United Kingdom; (right) in Guatemala.

Some of the 'Forest and woodland' category in Table 4.2 is the true climax vegetation of the area concerned, but in many cases is actively managed as plantation or other forestry: see Figure 4.4 for examples. The 'Other land' category includes those areas with relatively little vegetation – deserts, tundra, rock and ice-fields, urban land and that occupied by mineral workings.

Activity 4.2

(a) On the basis of the descriptions given above and in Table 4.2, what percentage of total land used is agricultural?

(b) Does the table indicate what percentage of the total is occupied by 'natural' vegetation?

The answers to the activities are given at the end of the chapter.

Table 4.2 shows that agriculture is the largest single use of land by humans, and on a global scale agricultural land use can be classified as in Figure 4.5. This is a very broadbrush picture, and does not tell us exactly what land use occurs in a particular area. There are two reasons for this. First, the map only shows agricultural activity, not the other uses in Table 4.2. Second, there is the question of scale, as will be clear if we consider the British Isles.

○   What is the agricultural land use to which the British Isles are allocated? Does this agree with your general knowledge of these islands?

●   The British Isles are mostly shown within the 'commercial dairy farming' region. While dairy farming is important, it is certainly not the dominant land use, which varies between different areas. Other major uses are arable production and rough grazing, the latter often on land that is regarded as having 'natural' vegetation such as grass and heather moor.

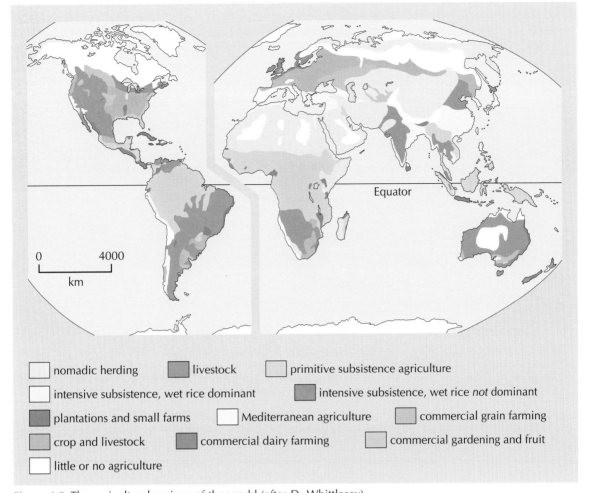

nomadic herding   livestock   primitive subsistence agriculture

intensive subsistence, wet rice dominant   intensive subsistence, wet rice *not* dominant

plantations and small farms   Mediterranean agriculture   commercial grain farming

crop and livestock   commercial dairy farming   commercial gardening and fruit

little or no agriculture

Figure 4.5 The agricultural regions of the world (after D. Whittlesey).

From Chapter One you should recall that the expected climax vegetation of the British Isles is temperate woodland. If you compare Figure 4.5 with the map of the world's biomes in that chapter (Figure 1.15), you will see that much of the agricultural land of the world occupies what would be expected to be forest biomes. Agriculture obviously changes the vegetation of an area, and in recent years there has been widespread concern that continuing conversion of forest land to agriculture may have serious consequences for environments. We will consider in the next section whether this concern is justified.

3.1 Change in land: deforestation

How we define forest or woodland is actually a matter of considerable debate. A recent study by the Food and Agriculture Organisation (FAO) of the United Nations reported that different authorities define forest using as a criterion anything between 10 and 20 per cent of the land surface covered by tree canopies, depending on what other use is made of the land. Table 4.3 gives original FAO estimates of forest and woodland area in different regions over the period from 1961 to 1991, and their greatly modified data for 2000. As a result of changes in definition and in methods of gathering data, the figures for 2000, taken from the FAO *Forest Resources Assessment 2000*, are not always consistent with data from their statistics database for earlier years. In particular, the figures for the EU and for North and Central America for 2000 cannot realistically be compared to those for the same named areas in earlier years.

Table 4.3   The percentage of forested land in different regions, 1961–2000

	1961	1971	1981	1991	2000
Africa	24.2	24.1	24.0	23.6	22.44
Asia (excluding USSR)	21.4	20.8	19.9	19.4	18.26
European Union (12)[1]	20.5	22.7	23.8	24.6	46
North and Central America	37.1	36.6	35.5	36.4	26
South America	53.2	51.5	51.9	52.7	51
Oceania	23.4	23.4	23.4	23.4	23
USSR/Former USSR	40.8	40.8	41.4	42.0	n/a
World	32.6	32.2	32.0	32.1	30

Note: [1]The 12 member-states of the EU during the 1980s.
Source: Data from FAOSTAT.

Activity 4.3

Summarize the main variations in forest area over time and space indicated by Table 4.3. What overall change in forested area does the table suggest has occurred over the period shown?

Table 4.3 suggests considerable confusion over the changes in forest area, but is this a major environmental problem? Much of the concern centres on loss of moist tropical forests, mainly from logging or conversion to farmland, although there have also been losses associated with mineral extraction and urban development. Some authors claimed that the tropical moist forests in some areas would 'have disappeared completely by 2000' (Gomez-Pompa et al., 1972). Loss of forest in some areas has been spectacular, even if the total area involved is not a large percentage of the global total (see Figure 4.6).

○   Suggest some reasons why loss of tropical moist forest might be regarded as particularly serious?

●   There are probably three major reasons. First, these areas have high biodiversity (Chapter One, and **Freeland 2003**). Second, if the trees are burned to clear the land or allowed to rot naturally after felling, this releases a large amount of the greenhouse gas,

Figure 4.6 Forest fire and associated smoke pollution in Indonesia in 1999. It is suggested that many of the fires were started deliberately to clear the forest for other uses, but then went out of control.

carbon dioxide, into the atmosphere. Third, as mentioned in Chapter One, most of the plant nutrients in the forest ecosystem are bound up in living material and are easily lost through leaching if the vegetation is disturbed.

The concern over deforestation raises an important question of measurement. The data quoted in Table 4.3 appeared consistent up to 1991, then completely awry in 2000. The first problem is one of definition, as noted above (and see Figure 4.9). The second is accessibility. Much of the area occupied by tropical moist forest is, almost by definition, difficult to traverse, so that accurate maps are difficult to compile from ground level. This is an especial problem for poorer countries that lack the funds to undertake detailed surveys. Even where this is possible, there may be all sorts of reasons why a particular country might wish to exaggerate or to play down changes in forest area. Given all these uncertainties, some authors have claimed that, rather than there being any danger of their disappearing, 'conversion of forested areas is a rather modest and localized process' (Sedjo and Clawson, 1983).

Is it possible to resolve this? A partial answer is offered by remote sensing (see Box 4.1). Using data from a range of such studies, FAO concluded that their original estimates – that 11.3 Mha of tropical forest had been lost per annum between 1981 and 1990 – should be raised to 15.4 Mha per annum (Freeman and Fox, 1994). By 2001, using more data, they quoted a loss of 8.6 Mha per annum between 1990 and 2000.

Box 4.1 Remote sensing

Since the 1920s, and especially since the development of satellites and telecommunications in the 1970s, data about the world's surface features have increasingly been gathered remotely. Aircraft and satellites can overfly large areas rapidly and, originally using photography but now using a variety of related technologies, take 'pictures' of the surface below. At their simplest, these could be conventional monochrome photographs that could be interpreted and measured by humans to produce maps of the vegetation or other features of the area studied. More recently, colour photography has provided additional information for the photo-interpreter. (See Figure 4.7.)

Fundamentally, all remote sensing involves measuring the radiation reflected from a point on the surface, and recording that measurement either directly on a photograph as a shade of grey or a colour or as a number in a computer. Photographs only record reflected light in the visible, or near visible, ranges of the spectrum. The advent of sensors that could measure reflected radiation over a wider range, and computer software that could manipulate the data, have made remote sensing a much more powerful technique. The size of the smallest 'point' that can be recorded has decreased, and the frequency with which satellites overfly each area has increased enormously. Potentially, 'images' of almost any point on the Earth's surface, down to a resolution of a few metres, are available almost on a weekly basis.

However, we need to remember what these 'images' show. Associated with each point in the image there may be some 5 or more numbers for the radiation received by the sensors at different wavelengths. As raw numbers, they are pretty meaningless, and there are all sorts of difficulties with the geometric registration of adjacent or subsequent scans over an area. To make sense of these data requires several things. First, there need to be data about the vegetation (or whatever is of interest) measured on the ground for at least a sample of points in the area involved. Then, a human, or more likely a computer, has to compare the image data with the 'ground truth' data, and work out what combinations of data correspond most closely to a particular aspect of ground truth. This is an enormously powerful technique, but there is always going to be some uncertainty. The 'ground truth' data can only be a **sample**, and possibly not a completely representative one, of the area of interest. The correspondence between the combinations of remotely sensed data and the ground truth is also unlikely to be exact. However, it does give us much more extensive and reliable information than could possibly be gained any other way. (See Figure 4.8.)

(a)   (b)

Figure 4.7 Aerial view of Milton Keynes: (a) 'true' colour image; (b) false colour.

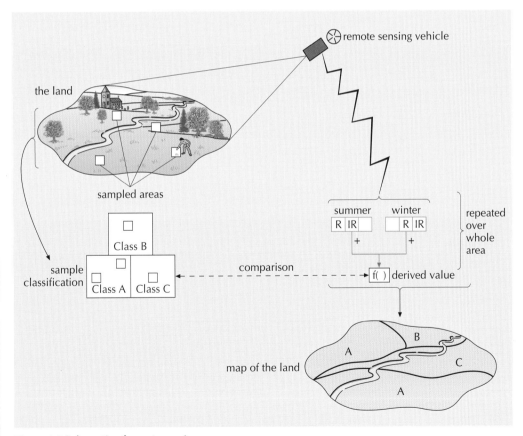

Figure 4.8 Schematic of remote sensing process.

Figure 4.9 Different forms of vegetation with trees. How do we decide what is forest?

○   The total area of forest and woodland in the tropical regions in 2000 was 1571 Mha. Calculate the annual rate of loss of tropical forest over the 1990–2000 period quoted by FAO as a percentage relative to this total area.

●   8.6/1571 = 0.55 per cent per year of total forest area across the tropics.

The earlier columns of Table 4.3 suggested very little change in the area of forest and woodland in Africa and a decrease in the area in South America. The FAO study in 1990 suggested that losses of tropical forest were greater than estimated, then the most recent data appear to suggest it had now decreased. In part, this can be explained by the problems in gathering and collating data, but it also conceals the fact that while the rate of complete loss may have decreased, the nature of the forest can still change significantly. Even if an area is still forested, according to some agreed definition, there is a lot of difference between the original highly diverse tropical forest and the **secondary growth** forest that may regrow spontaneously after clearing or the managed forest that may be deliberately planted. These secondary forests are unlikely to provide the same diversity of structure and hence of species as would the original undisturbed forest. The change in forest type may also be associated with other possibly damaging human activities, such as increased use of pesticides or the introduction of alien species. So while the net loss of *forested area* may not be great, the environmental impact of the specific loss of climax tropical moist forest may be very significant.

This discussion of deforestation illustrates our key concepts – space and time, and uncertainty. What looks very serious at one scale (as in the Indonesia example of Figure 4.6) may be less important relative to the whole global picture. The actual data about forest loss are difficult to obtain, and the data can be presented and interpreted in different ways. Compare the apparently stable situation that is implied by Table 4.3 with the more alarming data from Freeman and Fox (1994)

secondary growth

cited on page 147 above. This uncertainty adds to the controversy over change in environments.

Summary

- Globally, land can be categorized according to the expected climax community that it would support, but a large proportion of the land surface has been modified by human activity.
- Historically, humans have converted a major proportion of the forested area of the world to agricultural use.
- Current rates of change in forest area are a matter of controversy, but are probably high enough to cause concern.
- Use of remote sensing provides one possible way of decreasing the uncertainty over change in land.

4   Below the surface: land and soil

The nature of 'land' is not just defined by its surface. We need also to look below the surface, particularly at the soil, where many essential ecosystem processes take place, as outlined in Chapter One. Perhaps surprisingly, it is difficult to give a rigorous definition of soil, other than the commonsensical one that it is the thin uppermost surface layer of the Earth, distinct from the underlying solid rock. Many of the contrasts in vegetation that we see around us are the outcome of processes in the soil that reflect differences in these underlying rock types. One example is the contrast between the birch/pine/heather community and the beech/oak woodlands that often both occur relatively close together in northern Europe, but on soils overlying granite/sandstone or limestone/chalk respectively (see Figure 4.10).

Soils contain a vast range of living organisms, from earthworms to bacteria, and represent an enormous, though so far poorly documented, proportion of total biodiversity **(Freeland, 2003)**. Human activity can have devastating effects on

Figure 4.10 Two different communities, mainly reflecting differences in the underlying soil: (left) the birch (*Betula* spp.) community; (right) beech (*Fagus* spp.) woodlands.

Figure 4.11 (left) A farm swallowed up by windblown soil: erosion in 1930s' USA resulted in the 'Dust Bowl'; and (right) farming on inappropriate land, such as this steep slope in Uganda, is starting to cause soil erosion.

soil erosion

soil, and hence on our environment. During the 1920s and 1930s areas of the USA that had been used primarily for arable cultivation suffered severe **soil erosion**, the infamous 'Dust Bowl': see Figure 4.11(left); similar problems can follow deforestation, as Figure 4.11(right) shows. To anticipate and avoid such problems and to understand other changes in environments, we need to examine processes that occur in soils, some on a microscopic scale.

pores

mineral
organic

clay
silt
sand
gravel
texture

A typical soil consists of a mixture of solid particles, water and air. The water and air occur in the gaps or **pores** within the solid part of the soil, which is itself made up of different **minerals** and **organic** (carbon-containing) materials. The minerals are derived originally from the solid rock and the organic matter from living organisms. The mineral particles are classified by size as **clays**, **silts**, **sands** or **gravels**: see Table 4.4. Different soils have different mixtures of these particles, which define the **texture** of the soil.

Table 4.4   The range of different soil particle sizes

Particle type	Clays	Silts	Fine sands	Coarse sand	Gravel
Size range (mm)	<0.002	0.002 – 0.02	0.02 – 0.2	0.2 – 2.0	>2.0

○   From your experience, how would you say the properties of a soil that is mainly clay differ from one that is mainly sand?

●   Clay soils form a smooth, sticky mass when wet and a hard, solid mass when dry, while sandy soils are much more crumbly under all conditions.

The basic texture of a soil is only part of this story, since the smallest clay and silt particles can be bound together into larger units – or **soil aggregates** (see Figure 4.12) – to give the soil a **structure,** which affects its air and moisture content. Most soils have a network of large pores (>0.5mm) between the aggregates, combined with smaller pores within the aggregates themselves. Water and air can flow freely through the larger pores, but the smaller ones act like blotting-paper, and prevent the water in them from draining away. The 'blotting-paper' forces are called **capillary forces** and the smaller the pore, the stronger is the force holding the water in place.

soil aggregate

structure

capillary force

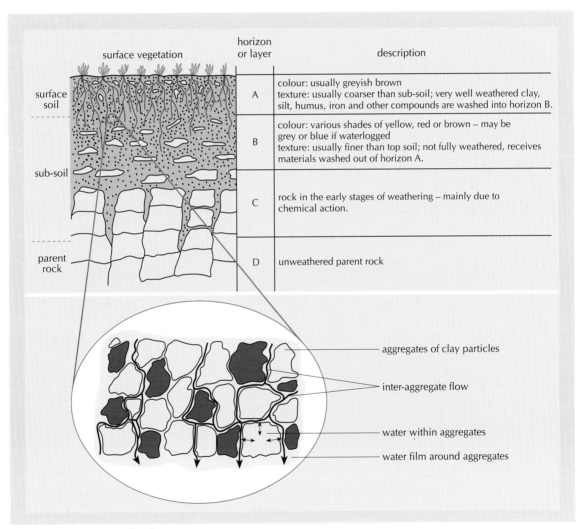

Figure 4.12 Diagram of soil structure.

The 'best' soils from a human perspective have a mixture of pore sizes. The network of large pores lets any excess rainfall drain away and allows air to reach respiring plant roots. Roots can overcome the capillary forces to extract water from the finer pores. An appropriate mixture of large and small pores means there is a store of water in the soil that plants can draw on to grow, but not so much that it is waterlogged. This ideal structure occurs naturally in soils under temperate woodlands, but gardeners will be familiar with the effort that can be needed to create it elsewhere!

soil formation
parent material
weathering

The soil on any given site results from continuously occurring **soil formation** processes: see Figure 4.13. The **parent material** (either solid rock or older soil) is broken down into smaller particles by various processes of **weathering**. The daily and annual cycles of heating and cooling of exposed surfaces lead to continual expansion and contraction that can crack the parent material. In higher latitudes, water near the surface of a rock can freeze, expanding as it does and exerting a force of around 150 kg on every square centimetre to shatter it (think of burst pipes!). Moving water breaks particles off the surface, and dissolves out soluble materials. This is aided by the **acidity** of the water caused by excretory products from living organisms or carbon-, nitrogen- or sulphur-oxides absorbed from the air. Where the rock or other parent material has a high content of calcium (as in chalk or limestone), this is readily dissolved out by acid water, neutralizing the acidity to provide a **neutral** or **alkaline** soil.

Acidity will be discussed in more detail in Chapter Six.

neutral
alkaline

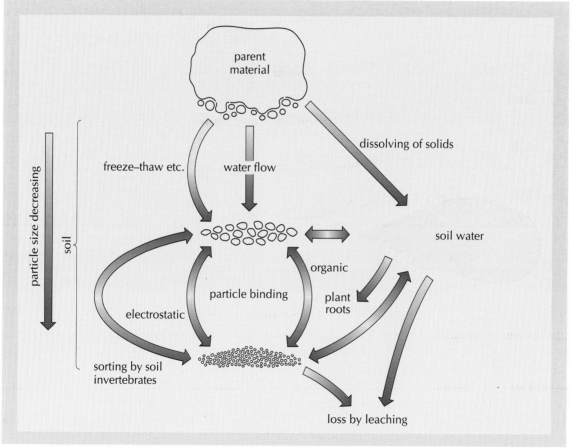

Figure 4.13 Soil formation processes.

Some plants such as scabious (*Scabiosa* spp.: see Figure 4.14a) grow best on neutral or alkaline soils; others such as heathers (*Calluna vulgaris*) and mat grass (*Nardus stricta*) will grow, albeit generally slowly, on the acid soils formed from calcium-poor rocks such as granite and some sandstones: see Figure 4.14(b) and (c). Similarly, animals like snails, with shells that have a high content of calcium, survive better on more calcium-rich, neutral or alkaline soils. The nature of the soil thus directly affects the living community (Figure 4.10) but the community also influences the soil. Roots can exert enormous pressure that can break apart solid particles, and secretions from roots increase the rate of weathering.

(a)

(b)

(c)

Figure 4.14 (a) The Devil's Bit scabious (*Scabiosa succisa*), typical of alkaline soils; (b) heather (*Calluna vulgaris*) and (c) mat grass (*Nardus stricta*), both typical of acid soils.

Weathering processes tend to produce smaller and smaller particles, but this effect is counterbalanced by other effects that bind them together to give the soil its structure. Processes at the molecular level produce electrical charges on the surface of the smallest clay particles (an effect like the electrical shock you sometimes get from clothing made of artificial fibres). These electrical charges produce forces that tend to bind the smallest particles back together. This gives rise to the solidity of clay soils but is only effective over microscopic distances, so electrical forces between sand-sized particles are negligible. Soil particles are also held together by gums and other products secreted by plant roots or produced by detritivores and decomposers involved in the decay of living organisms.

humus

The more resistant parts of plant debris may not be broken down at all, especially under wet, poorly aerated conditions where they form a layer of peat. Peaty soils are typical of high latitudes and are usually acid, with a distinctive plant community of heathers, sedges and grasses. Under better aerated conditions, the organic remains partially break down to form **humus**, a brown, relatively uniform material which also helps to stabilize soil structure. Water percolating down through the soil moves materials with it, while soil animals such as worms can carry them back towards the surface, and their burrows create additional structure. Chemical reactions involving different forms of iron, calcium and other elements give rise to different coloured compounds, usually red or yellow in well aerated soils or blue-grey (gleyed) in waterlogged ones. As materials are sorted, moved up and down the **soil profile**, and deposited again, distinguishable layers in the soil called **horizons** are produced: see Figure 4.15. Different sets of horizons occur in the soils of different climatic regions and parent materials. Some of the most distinctive soil horizons occur on heather moorland, and are visible when the soil is exposed on the edge of road cuttings, quarries and so on. Where soils have been cultivated for crops, the different horizons are usually mixed up and difficult to distinguish.

soil profile
horizon

Soils are thus continually changing over time, through physical and chemical processes as well as those involving living organisms. The organic matter and soil organisms in particular play a crucial role in storage and release of plant nutrients, with major effects on the whole living community, as we saw in Chapter One. Nutrients occur in three main forms in the soil: in the tissues of plants, animals etc.; as individual molecular-scale particles in solution in the soil water; and as insoluble mineral components. The three major nutrients – nitrogen, phosphorus and potassium (the N, P and K of garden fertilizers) – behave in different ways in the soil, and this has important implications for environmental change. As with the small clay particles considered above, individual molecules of many substances can acquire a positive or negative electrical charge. Potassium (K) dissolved in water can form a positively charged (K^+) **cation** (pronounced cat-eye-on). Nitrogen (N) can combine with hydrogen to form a positively charged ammonium cation (NH_4^+) or with oxygen to form negatively charged nitrate or nitrite **anions** (NO_3^- or NO_2^- respectively). Phosphorus (P) occurs combined with oxygen as the negatively charged phosphate ion PO_4^{2-} but also combines chemically with other elements to form insoluble compounds that become part of the soil aggregates.

Chemical terminology was introduced in Box 1.1 and in Section 3.4 of Chapter Two.

cation

anion

The smallest (clay) particles in soil have a net negative charge. Positively charged particles, such as the cations in the water within the pores in the soil, are attracted towards negatively charged materials like the clay surfaces. Negatively charged particles, such as nitrate ions, are repelled by the negatively charged clays, so move to the parts of the soil water further away from them (see Figure 4.16 overleaf). Different clay minerals have different amounts of negative charge per unit mass, and so can hold different amounts of positive ions in the soil water near their surface. This is called their **cation exchange capacity**.

cation exchange capacity

(a)

(b)

(c)

(d)

Figure 4.15 The profiles of some different soil types: (a) podzol (with organic matter above and iron-rich layer below); (b) rendzina (showing thin soil layer over chalk bedrock); (c) gleyed soil on alluvial clay; (d) brown soil with gleying beneath.

○   Which of the ions named earlier would you expect to find in the soil water close to the clay surfaces?

●   The positive ions, potassium and ammonium.

The phosphorus that is chemically bonded into the soil aggregates is least readily washed out, or **leached**, by water flow. Water moving through the pores can overcome the weaker electrical forces attracting cations to the clay surfaces, so these ions are removed gradually by leaching. Negatively charged ions like nitrate that are repelled by the clay surfaces are most easily leached.

leaching

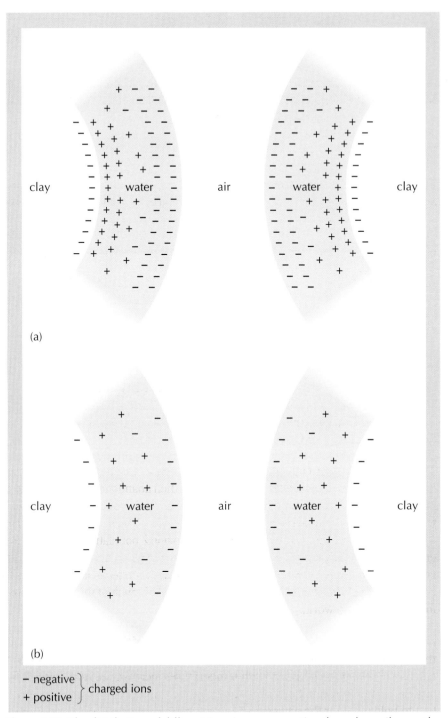

Figure 4.16 The distribution of different ions in a cross-section through a soil pore: the example in (a) has a high cation exchange capacity, and in (b) has a lower one.

The dynamics of nutrients in the soil at this microscopic scale have important field-scale effects, especially in intensively farmed areas and the tropical moist forests. A gardener or farmer can replace leached nutrients using manufactured fertilizers (at a price) and the decomposers in the soil replenish them as they break down plant materials in an undisturbed community. Environmental changes occur when gains and losses of nutrients in the soil are not equal.

As we shall see in the next section and in Chapter Five, where fertilizers are used to supplement soil nutrient, any surplus that is not taken up by plants can be leached out of the soil, into local surface or underground water and ultimately into the oceans. The presence of increased nutrients in water can cause a number of problems, as we shall see in the next chapter.

The lush vegetation of the tropical moist forests suggests very 'fertile' conditions, apparently attractive for conversion to agriculture by removing the existing vegetation and either burning it or allowing the softer material to rot and be incorporated into the soil. If the vegetation is burned, this releases the potassium and phosphorus stored in the vegetation immediately into the soil, although the nitrogen is lost as oxides into the air. The nutrients that are released allow rapid crop growth in the first year after forest destruction. However, the high rainfall in such areas, that allowed the formation of forests, means that this is not a sustainable system.

○   Why might such a system of cultivation not be sustainable?

●   Rainwater flowing down through the soil will carry away nutrients such as nitrate. Together with the removal of nutrients in the crop for human use, this rapidly depletes nutrient supplies, so crop yields decline after only a few years.

This **slash-and-burn** cultivation can be sustained at low population densities, ~~slash-and-burn~~ because areas are cultivated for a short time but then abandoned. Over a period of 20 or more years, the abandoned areas will undergo normal succession and a forest community reappears, accumulating nutrients. This secondary forest can then be cut down, and the cycle repeated. Provided population density is low enough for the cultivators to survive on the product of about one-tenth of the total area, they can use each patch for a couple of years, then leave it for 20 years to recover. Such **shifting cultivation** has been typical of much of the tropical moist ~~shifting cultivation~~ forest areas of the world.

Nutrient losses are only part of the story. The removal of the forest cover exposes the soil to the direct impact of rain, and this can break down its surface structure and carry loose material away. High temperatures increase the rate at which the organic matter in the soil is broken down by decomposer organisms, so there is less to hold together the solid particles, making them easier to detach and erode away.

Similar mechanisms explain the 'Dust Bowl' of the USA where, prior to European invasion, the prairie soils were in a relatively stable condition subject to grazing by wild animals. The dung of these animals, and the uneaten grass that died off each year, kept the soil organic matter and plant nutrient supply in a steady state. Ploughing the prairies allowed the European settlers to grow successful grain crops, using the nutrients released from the decaying organic matter, plus small amounts of fertilizer. However, the initial ploughing removed the protective effect of the vegetation and increased the rate of decomposition of organic matter exposed to the sun and air. The removal of grain and straw to the feedlots or to urban parts of the USA reduced the local return of organic matter to the soil. Hence, over time, both soil nutrient supplies and the structure of the soil were degraded, making the soil susceptible to severe erosion.

During the latter half of the twentieth century, it was claimed that a similar situation was arising in the UK and parts of Europe, where arable crops were apparently being grown in a way analogous to 1930s' USA. There were some spectacular, and many less spectacular, examples of soil erosion in areas where it had not been seen before (Figure 4.17, left). The pressure exerted on the soil by some agricultural machinery is higher than that permitted for lorries on roads that are specifically designed to carry them. The increased use of such larger machinery, causing compaction of the soil structure, was also cited as a potential cause of erosion (Figure 4.17, right).

Undoubtedly, there were local problems at this time, but the situation was actually different from that in the USA in three important respects. These were the more uniform rainfall with less extreme events, lower summer temperatures and the greater use of fertilizers in Europe compared to the USA. The lower temperature reduced losses of organic matter, and the less extreme rainfall produced less surface movement of soil particles.

○   How might the use of fertilizers affect soil organic matter?

●   Fertilizers increased the growth of the crop, including the roots and unharvested stubble that were left behind. Incorporating these residues into the soil helps to maintain levels of organic matter.

Figure 4.17 (Left) Rill gully erosion in a field under cultivation in Europe; (right) a sugar-beet harvester, typifying the heavy agricultural machinery in use in modern farming.

Figure 4.18 shows the organic matter levels in soils that were cropped for wheat every year in the UK, with or without fertilizers, or with farmyard manure (a mixture of decomposed straw and animal manure).

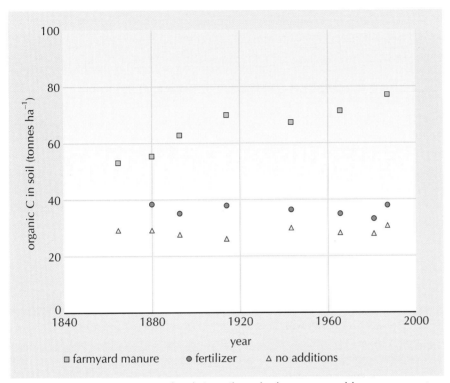

Figure 4.18 Organic matter (C) levels in soils under long-term arable management. *Source:* Jenkinson, 1990.

Activity 4.4

Do the data in Figure 4.18 suggest that organic matter levels can be maintained under continuous cereal cropping in the UK?

This discussion emphasizes again the importance of examining environmental processes at a range of spatial and time scales. Biochemical processes that take place within the soil at microscopic scale and soil formation processes that take many decades or longer can change the living communities on those soils on different scales. The behaviour of soils when cultivated also varies, depending on their type, spatial location and the precise forms of cultivation employed. Movement of materials across space by human activity can also affect soils.

Summary

- Soils are the medium in which many important environmental processes occur, and they can affect directly or indirectly a whole range of other features.

- Soils are not inert but are constantly changing as a result of physical, chemical and biological processes.
- The chemical behaviour of plant nutrients on a microscopic scale in the soil can influence both the immediate and more distant living communities.
- Human actions can affect soil properties, and hence other environmental processes that depend on them.

5   Agriculture and environment

We have already noted that agriculture is a major use of land worldwide, changing vegetation from its climax form and possibly causing major change in soils. This section looks at some other environmental effects of agriculture, and tries to balance these against its essential purpose of supplying food for humans. Successful food production depends on converting as much as possible of potential primary production into appropriate foodstuffs, ideally without reducing the long-term production potential.

○   What are the major factors affecting primary production?

●   Production depends on the solar energy income and the proportion of this trapped by plants. That depends on how much leaf they have, but growth can also be limited by the supply of water and soluble plant nutrients from the soil, and by temperature.

The increase in human population, documented in Chapters Two and Three, suggests that agriculture has been largely successful, particularly since the eighteenth century, in increasing food production. The preceding chapter gave some data for the change in food supply, as recorded by the United Nations Food and Agriculture Organisation. Given some of the doubts associated with the forest cover data presented in Section 3.1 (Table 4.3), you may be dubious about these data, but the general picture presented in Figure 4.19 is probably correct.

Activity 4.5

(a) Looking at Figure 4.19, which region shows the greatest percentage increase in total food production between 1961 and 2001?

(b) What does Figure 4.19 suggest with regard to availability of food per capita over the period?

(c) How did European food production change over the 1961–2001 period? Suggest two possible contrasting explanations for these changes in European production.

Production of food appears to have increased in a quite remarkable manner over most of the period. However, the global picture conceals wide variation between

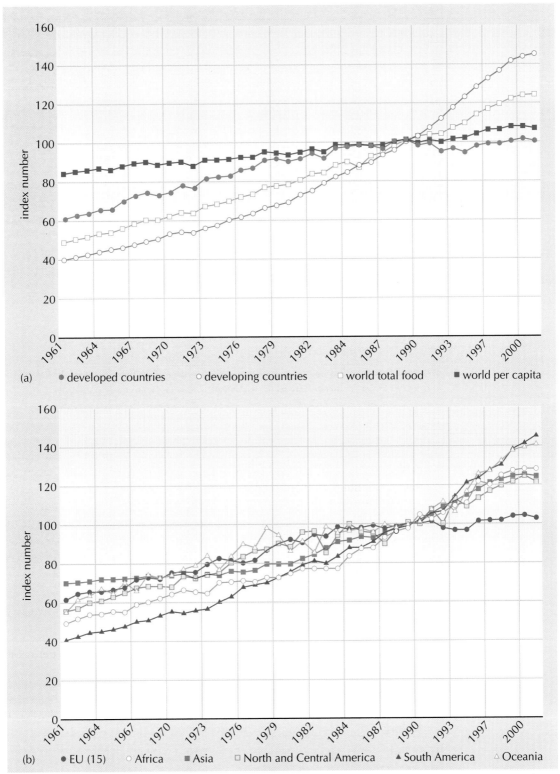

Figure 4.19 (a) Total food production for the developed world, developing world and whole world, and per capita production for the whole world, 1961–2002; (b) total food production 1961–2002 for different areas. *Notes:* All data are from FAO statistics, and are presented relative to a baseline of 100 for 1989–1991. EU (15) refers to the 15 countries that were member-states of the European Union in 2000.

areas, and this success has not been achieved without costs, both financial and environmental. Over recent years concerns have grown over agriculture's effect on other aspects of the environment, first called into question by Rachel Carson in her book *Silent Spring* (1962):

> Over increasingly large areas of the US, spring now comes unheralded by the return of the birds, and the early mornings are strangely silent where once they were filled with the beauty of bird song.
>
> (Carson, 1962)

pesticide

Carson largely blamed this on the use of **pesticides**, synthesized compounds used to control undesired species of pests, weeds or disease in agricultural crops and livestock, but subsequently agriculture has attracted other criticisms. We will examine some of these in the following sections.

5.1 Environmental effects of fertilizer use on land

It is generally recognized that the use of fertilizers has been a major factor contributing to the increase in yields noted in Figure 4.19. However, any excess fertilizer leached out of the soil may cause damage to water courses, as we shall see in the next chapter. There is one other potential environmental impact from the use of fertilizers, and that is the need for energy (usually in the form of fossil fuel) to produce the fertilizers. Fuel is needed to mine the raw materials for some fertilizers. The chemical reactions involved in converting nitrogen from the air into soluble forms for use as fertilizers are highly endothermic (see Chapter Two). The root nodule bacteria found in plants such as clovers can achieve this conversion at ambient temperature, obtaining their energy from the carbohydrate produced by the plant's photosynthesis. However, to achieve the same result industrial processes need high temperatures and pressures and large amounts of fossil fuel.

Phosphorus and potassium for fertilizers are mined from deposits found on land in a few areas such as Saskatchewan in Canada, North Yorkshire, parts of the former Soviet Union, as well as from offshore sources. The amount of fossil fuel required to mine, process, transport and apply phosphate- and potassium-containing fertilizers is about 14 and 9 MJ per kilogram of nutrient respectively (Leach, 1976). However, Leach estimated that nitrogen-containing fertilizers require some 62–84 MJ (equivalent to about two litres of oil) for each kilogram of nitrogen. Since a typical UK cereal crop receives over 170 kg of nitrogen per hectare, that requires some 350 litres of oil. In comparison, the possibly more obvious use of fuel to cultivate, sow and harvest the product from that area would be some 40 litres (Leach, 1976). Current world production of fertilizers represents 80 M tonnes of nitrogen per year.

Activity 4.6

(a) On the basis of the behaviour of different nutrient elements in the soil, is the high energy cost of producing nitrogen-containing fertilizers likely to be more or less serious than the energy cost of those containing phosphorus or potassium?

(b) Estimate the total global energy cost of producing nitrogen-containing fertilizers.

5.2 Irrigation and salinization

In the drier regions of the world, maximizing crop production often relies on supplementing its water supply by **irrigation**. The next chapter will consider the availability of this water, but it is important to note that irrigation can in the long term have detrimental effects on the land and its ability to be used for agriculture.

When a crop is irrigated (Figure 4.20), some water is immediately taken up by the plant roots, some remains in the soil and some will flow down to lower horizons. If irrigation is then discontinued on the assumption that the plants have been adequately supplied, evaporation from the surface will effectively draw water back up from below. If the lower layers contain deposited salts such as sodium chloride, these salts are carried up with the water and concentrated on the surface as it evaporates into the atmosphere.

Figure 4.20 Irrigation equipment: sprinklers irrigating a field of potatoes in Colorado, USA.

salinization

Further irrigation should wash the salts back down, but in order to ensure that this happens, enough water has to be applied to produce a net downward movement, and this may be more than is apparently needed for the crop to grow well in that season. The accumulation of salts near the surface causes **salinization** which can prevent normal crops, or the 'natural' vegetation of the region, growing there: see Figure 4.21. Although apparently a good thing, irrigation can therefore be seen to be possibly detrimental in the long run by making the land unusable for agriculture or requiring saline-tolerant crops. This problem has affected major areas of Australia, the USA and elsewhere.

Figure 4.21 Example of salinized soil in a sugar cane field, Tanzania.

5.3 Agriculture and biodiversity

monoculture

The pesticides currently used are very effective at their job, so that many crops are now nearly complete **monocultures**, that is given over to just one species. The virtual elimination of weeds has also reduced or eliminated some of the insect species that feed on them. Insects, weeds and in some cases the stubbles left after grain harvest are also a source of food for farmland birds whose numbers in the UK declined dramatically during the 1980s and '90s (Fuller et al., 1995). It is important to recognize that the species concerned are defined as *farmland* birds (see marginal drawings for examples). That is, they are characteristic of a pattern of land use resulting from a specific set of historic changes as documented in Chapter Three. This means that communities of organisms on cropped land are very different from those typical of the local biome. It is therefore possible to interpret recent falls in population as just the latest of a continuing series of changes.

corn bunting (*Miliaria calendra*)

Land in Europe has been largely dominated by agriculture for many centuries, and many of the species used in agriculture were found there anyway, albeit not in monocultures or so ubiquitously. In contrast, Australia at the time of European settlement had a distinctive endemic flora and fauna (as you will remember from Chapter One). The introduction of completely new grazing species – sheep and cattle – and the escape of other non-native species such as rats and rabbits, coupled with a failure to understand the dynamics of the Australian ecosystem, have created major changes in the land there (see Figure 4.22, and **Freeland, 2003**).

linnet (*Carduelis cannabina*)

skylark (*Alauda arvensis*)

tree sparrow (*Passer montanus*)

lapwing (*Vanellus vanellus*)

genetically modified organisms (GMOs)

Figure 4.22 The effects of the introduction of European-style agriculture on Australia: the patches of forest left by the clearers in this area were doomed and could not survive.

Alien species can have enormous effects on the native communities of an area. This is a concern with the introduction of **genetically modified organisms (GMOs)** as agricultural crops (examined in **Bingham, 2003**). Such crops may be considered legally as no different from their unmodified counterparts, but even if a modified crop is carefully engineered not to become invasive, it is possible that if it did interbreed with closely related native plants the result could be 'superweeds' that disrupt existing species and communities.

5.4 Agriculture as hero or villain?

Agriculture can clearly have major environmental effects, and it occupies much of the world's land. Has its influence, taken as a whole, been for good or ill? In 1993 Pretty and Howes published a table summarizing what they regarded as the agricultural intensification

principal environmental and health problems caused by **agricultural intensification**. The phrase covers a whole range of aspects, including the increased use of pesticides and fertilizers, but also increases in mechanization, numbers of livestock kept on a given area and indoor livestock-rearing. Their summary is reproduced as Table 4.5.

Table 4.5 appears at first sight to present a damning indictment of agricultural intensification. Many of the items in the list represent direct, although often contested, threats to human health. If proven, then they certainly need to be minimized. An overall evaluation of environmental change resulting from agriculture is more difficult. Carson's dire prediction of 'no birds singing' has not yet been realized, despite the decline in the numbers of European farmland birds. So, do the observed changes matter? In the UK and Europe the current situation is often unfavourably compared to that which existed sometime between 1930 and 1960.

○   To what extent was the land surface of the UK occupied by 'natural' communities at that time?

●   From Figure 1.15 in Chapter One, the UK is in the temperate forest biome, so that would be its natural community. By 1930 most of the land was already used for agriculture or affected by it, so most communities were human-influenced.

The European land surface has been dominated by relatively homogeneous cultures for several hundred years and the changes that currently cause concern relate to a more or less arbitrary baseline. For the New Worlds of the Americas and, particularly, Australia, the advent of European settlers represented a radical, human-induced change from something much closer to a natural situation. This again stresses the importance of time in considering environmental change. Over what period has change occurred, and is it part of a continuing process or does it represent some sort of step change?

In Europe obtaining food is so simple that it is easy to forget that elsewhere its availability can be a problem. It may be fair to ask whether the importance of securing the supply of sufficient grain for the world's increasing population outweighs such things as the loss of what some would regard as 'a few obscure brownish birds'. I leave you to ponder that.

Activity 4.7

Summarize the main ways in which agriculture affects environments. How important are these effects?

The answer to this question provides a summary of Section 5, and you can compare your conclusions with those provided at the end of the chapter.

Table 4.5   The principal environmental and health problems caused by agricultural intensification in Britain

Contaminant/pollutant	Consequences
Contamination of water	
Pesticides	Contamination of rainfall, surface and groundwater, causing harm to wildlife and exceeding limits for drinking water
Nitrates	Methaemoglobinaemia[1] in infants, possible cause of cancers
Nitrates, phosphates	Algal growth and eutrophication, causing taste problems, surface water obstruction, fish kills and illness due to algal toxins
Soil	Disruption of water courses
Organic livestock wastes	Algal growth, plus deoxygenation of water and fish kills
Silage effluents	Deoxygenation of water and fish kills; nuisance
Contamination of food and fodder	
Pesticides	Pesticide residues in food
Nitrates	Increased nitrates in food; methaemoglobinaemia in livestock
Antibiotics	Antibiotic residues in meat
Contamination of farm and natural resources	
Pesticides	Harm to farmworkers and public; nuisance; harm to predator populations; harm to wildlife
Nitrates	Harm to wild plant communities
Ammonia from livestock	Disruption of plant communities; possible role in tree deaths
Soil and water run-off	Flooding on and off farm; damage to neighbouring housing and communities; nuisance
Metals from livestock wastes	Raised metal content in soils
Pathogens from livestock wastes	Harm to human and livestock health
Contamination of atmosphere	
Ammonia from livestock manures	Odour nuisance; plays role in acid rain production
Nitrous oxide from fertilizers	Plays role in ozone layer depletion and global climatic warming
Methane from livestock	Plays role in global climatic warming
Cereal straw burning [2]	Nuisance; enhances localized ozone pollution of troposphere; plays small role in acid rain production, ozone layer depletion and global climatic warming
Indoor contamination	
Ammonia, hydrogen sulphide from livestock	Harm to farmworker and animal health; odour nuisance
Nitrogen[3] dioxide in silos	Harm to farmworker health.

Notes: 1 Methaemoglobinaemia is commonly called 'blue baby syndrome'. The presence of nitrate and bacteria in water consumed by a baby affects its blood, reducing its capacity to carry oxygen, with possible fatal results.
2 A ban on straw-burning came into force in the UK at the end of 1992, as a consequence of the Environmental Protection Act 1990.
3 I think this was probably a misprint in the original table – carbon dioxide is the more serious threat in such circumstances.
Source: Pretty and Howes, 1993, Table 1.1, p.3.

6 Mineral resources and the environment

Humans also need land for dwellings, the infrastructure of roads, communication networks, power supplies, water supply and other industrial plant. To provide these requires fossil fuels, minerals and aggregates (the sand, gravel and similar materials used for construction) from beyond the immediate vicinity. There is a long list of such **resources** of useful materials required by humans and obtaining them can directly affect the land. Many of these resources are not uniformly distributed across the world but occur in greater or less concentrations depending on geological history and the processes of rock formation (outlined in Chapter One). Figure 4.23 shows the major sources of a range of important minerals.

The term **resource** is used here in the sense of Chapter Two, which is different from its use by Freeland and Drake in Chapter Three.

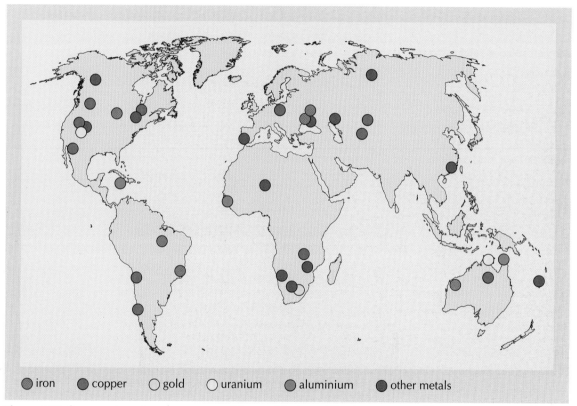

● iron   ● copper   ○ gold   ○ uranium   ● aluminium   ● other metals

Figure 4.23 Some major mineral resources of the world: sites of first importance with over 5 per cent of known world reserves.
Note: Other metals = manganese, mercury, tungsten, nickel, chromium.
Source: Adapted from *The Times Atlas of the World*, 1997, Plate 1.

○ From Figure 4.23, where are the major sources of copper located?

● USA, Chile, Canada, Zambia, Poland, Australia and the former USSR.

Every country has some land available for agriculture, although climate and history mean that its usefulness varies, but only a few countries have useful

sources of some minerals. This **unevenness** in the spatial availability affects the   unevenness
prosperity of nations, and also the impacts on their environments.

The use of different minerals by humans is basically determined by two factors:
their concentration in the near surface layers, and the ease of converting the raw
form into useful products. The first metals used by humans were those most
easily extracted (mixtures of copper and other metals to make bronze), but
accessible deposits of these were relatively rare. Iron occurs widely as oxides,
giving the orangey 'rust' colour to rocks and soils, and is relatively easily
extracted using limestone and elemental carbon such as charcoal. Iron was the
second major metal to be widely used. Local availability of limestone, wood for
charcoal, and then coal was a major factor in the birth of the industrial revolution
in Shropshire (Figure 4.24) and elsewhere. More recently, we have become
dependent on an increasing range of minerals, as alloying elements in
engineering metals, as catalysts for chemical processes and in semiconductors.

Figure 4.24 Early industrialization in the Ironbridge gorge, Shropshire: *Coalbrookdale by night* by Philippe Jacques de Loutherbourg, 1801.

Almost all minerals occur as compound chemical **ores,** mixed with other   ore
unwanted materials. The financial value of an ore source depends on how much of
the ore is present relative to the waste material. For minerals with a high value per
unit, it may be worthwhile to extract them from rocks containing only a fraction of
one per cent of the mineral, whereas low-value minerals, such as coal, may not be
worth extracting even from relatively rich deposits.

Mineral extraction causes several impacts on the land. There is the whole range
of buildings and other structures needed for extraction, processing and shipment,
and then the waste material or **spoil** needs to be disposed of (Figure 4.25). The   spoil

(a)

(b)

(c)

Figure 4.25 (a) An opencast coal mine, near Cologne, Germany; (b) Lee Moor china clay waste, Devon; (c) copper mine in Bingham Canyon, Utah, the largest human-made hole in the ground.

Bingham Canyon opencast copper mine in the USA, with its associated waste dumps, covers about 38 km² and is over 300 m deep. The ore extracted from the mine contained on average about 0.7 per cent copper, leaving a vast amount of waste to be dumped: see Figure 4.25(c).

The mineral extraction process often involves discharge of noxious gases as this is the easiest way to dispose of undesired material. The extraction and production of caustic soda in Lancashire in the eighteenth century was a notably polluting activity, causing the destruction of large areas of vegetation downwind of the factories. This led to the passing of the first effective pollution control legislation, the Alkali Act 1863.

Mineral extraction requires energy, with the potential for environmental damage that this can involve. Aluminium makes up more than 8.3 per cent of the Earth's crust, but it is currently only practicable to extract it from the mineral bauxite, a process which requires 210–374 MJ to produce one kilogram of metallic aluminium. About 1 per cent of total world energy supply is used for this aluminium extraction process.

A major concern in the early 1970s was that important mineral resources were about to 'run out'. At any time the **reserve** of a material is the estimated amount that could be extracted, given current technology and market price. Changes in

reserve

economics or technology can make it worthwhile to mine ore that has a lower concentration of mineral. A relationship called *Lasky's law* suggests that if it becomes economic to exploit rock with half the mineral concentration of that currently mined, this quadruples the reserve that is potentially available. So, paradoxically, the estimated reserve of a material can increase as more of it is used. Table 4.6 shows the estimated world reserves of four important metals in successive decades since the 1940s.

Table 4.6   Estimated world reserves of four metals, based on data from US Bureau of Mines

Period	Estimated reserves (million tonnes)			
	Copper	Lead	Zinc	Aluminium
1940s	91	31–45	54–70	1,605
1950s	124	45–54	77–86	3,224
1960s	280	86	106	11,600
1970s	543	157	240	22,700
1980s	566	120	295	23,200
1993	590	130	330	28,000

Source: Hodges, 1995, p.1306.

Activity 4.8

To what extent do the data in Table 4.6 support the idea that mineral resources are unlikely to 'run out' in the foreseeable future?

The argument in the answer to Activity 4.8 at least partially fails to consider the increased environmental impacts caused by the additional wastes generated and energy used to extract less concentrated ores. The total area involved in mineral extraction is relatively small (about 0.08 per cent of the UK land area, for example), but this does not mean that the popular response to the damage is limited. Limestone quarrying in the UK is one highly contentious case, since quarries tend to be situated in areas regarded as of high landscape value (remember Figure 4.1?). This effect is highlighted by the islands of New Caledonia which contain rich deposits of nickel, magnesium, chromium and manganese. The resulting local soils have high concentrations of these elements, and support a rich community of highly specialized organisms, able to tolerate these high concentrations though they would be unable to compete on normal soils (Flannery, 1994): see Figure 4.26. The very uniqueness of this vegetation increases the environmental impact of local mining operations.

In nineteenth-century USA, mining was regarded as the 'highest and best' use of land, irrespective of its suitability for other purposes, and current legislation in the USA is still based on the Mining Law of 1872 in which this principle is enshrined (Hodges, 1995). Hodges suggested that such attitudes were archaic in 1995, but at the beginning of the twenty-first century the US administration seemed well attuned to the views of the earlier frontier period.

Figure 4.26 The maquis vegetation growing on the red oxide soils of New Caledonia in the western Pacific.

Recently, the surface damage caused by mineral extraction has to some extent begun to be repaired and land reclaimed for other uses. In the UK, planning regulations require action to reinstate land after mineral workings, but this is often not the case in other countries. Table 4.7 shows the land area in England that was licensed for mineral extraction in 1988 and 1994 and the associated areas that had been restored between those dates. At that time, progress in restoration was still relatively limited for limestone quarrying and clay/shale extraction, but the record for coal and for sand and gravel was better.

Table 4.7   Permitted surface mineral workings in England 1988 and 1994 and areas reclaimed between these two dates (hectares)

Use	Permitted area 1988	Permitted area 1994	Area of land reclaimed 1988–1994
Sand and gravel	29,040	29,828	8,626
Limestone/dolomite	11,490	11,401	927
Clay and shale	10,090	9,107	852
Opencast coal	8,420	7,568	5,241
Deep-mined coal	n/a	n/a	1,692
Industrial sand	2,130	1,945	250
Ironstone	14,420	13,029	65
Total	75,590	71,128	17,653

Source: HMSO, 1994, Tables 2.1, 5.1.

There is one paradox associated with mineral working: this is illustrated by the following example. The solid wastes from the early alkali industry in Lancashire were tipped in mounds. Over time, these weathered to produce a highly alkaline soil, on which appeared plant species that were typical of chalk grassland, an environment completely different from the rest of the locality: see Figure 4.27. Time, through the processes of succession, has been a great healer and these tips have become highly regarded nature reserves!

Figure 4.27 Fragrant Orchid (*Gymnadenia conopsea*) growing on alkali waste.

Summary

- Humans require a range of mineral resources from land and acquiring these can cause major environmental change.
- The distribution of mineral ores is very uneven across the world.
- Changes in economics or technology can make it feasible to extract minerals from less concentrated sources, so that reserves are unlikely to limit human activity, but this may be at the expense of even greater environmental damage.
- Restoration of land after mineral extraction is feasible, but not always insisted upon.

Activity 4.9

Compare and contrast the environmental impacts of land use for mineral extraction with those of agriculture, in terms of their extent, unevenness and the actions that can be taken to mitigate them.

7   Total impact: our ecological footprint

ecological footprint

At the beginning of this chapter, I asked you to consider how many aspects of your lifestyle depended on the land. This same question has been formalized in the concept of the **ecological footprint** (Wackernagel and Rees, 1996), which tries to quantify the total human impact using land as the basic measure. The underlying concept is illustrated in Figure 4.28. An individual household, city or region is imagined to be placed in the centre of a hemisphere, which encloses a larger area of land. The area enclosed is calculated as that necessary to supply the needs of the chosen subject or, rather, of the humans associated with it.

Working in Vancouver, Wackernagel and Rees have devised a standardized method of calculating this 'footprint' area. They identify four requirements for land, three fairly obvious, while the fourth is conceptually more difficult and possibly controversial. A city, for example, physically occupies an area of land for buildings, roads, airports and so forth. The inhabitants of the city require food and water and this has to be supplied from agricultural land and the water catchment area. The buildings, and their contents require timber so an area of forested land is needed to supply this.

The fourth requirement for land relates to energy. Chapter Two showed that humans use a lot of energy to support their activities, and this has to come from the solar income or from fossil sources. If we had access to suitable technology,

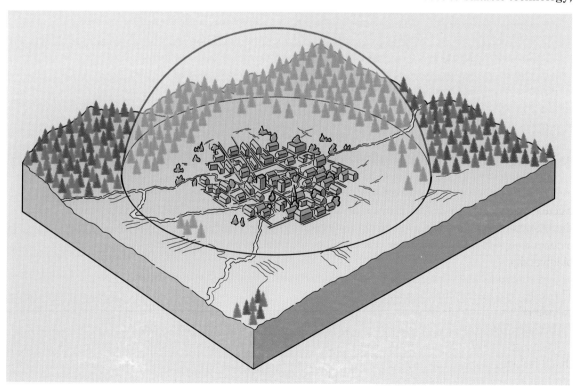

Figure 4.28 Diagrammatic representation of the area of land needed to supply all the needs of a city. *Source*: After Phil Testemale in Wackernagel and Rees, 1996, p.10.

then the energy supply for our city could be obtained by setting aside an additional area of land to collect solar energy. This land could not be used for agriculture or forestry, since these would compete for the incoming energy. **Blowers and Elliott (2003)** show that some energy could be obtained from photovoltaic devices on buildings whose area was already included in the footprint, but most of the energy needs of the 'developed' world are satisfied by fossil fuels.

One way of expressing this as a land area might be to total the area occupied by oil wells, mines, pipelines and related infrastructure, but this would still be a relatively insignificant amount of land, and thus would not adequately reflect its impact. Wackernagel and Rees use a combination of methods to quantify energy use in terms of land area, based on the growth of carbon-containing materials such as wood. They estimate the area needed to grow biomass fuels (wood, straw etc.) for those activities that actually used them. For fossil fuels, they calculate the area of forest that would be needed to absorb the carbon dioxide given off by their combustion. They postulate that for a city or other entity to be sustainable, it should not add to the carbon dioxide in the atmosphere.

○   Why would forest land be needed to absorb the carbon dioxide?

●   Recall from Chapter One that plants absorb carbon dioxide during photosynthesis, but also release it during respiration. Animals that eat plants also respire, releasing carbon dioxide, as do the decomposers that finally get hold of most living materials. The main medium- to long-term net store of photosynthetic carbon is in woody tissues, hence forests.

Of course, even timber does not last forever, so it is debatable how much 'energy land' a city needs so as not to add to atmospheric carbon dioxide. The calculation also makes no allowance for absorption in the oceans and other carbon storage processes (shown in Figure 1.8 in Chapter One and discussed further in Chapter Six).

Figure 4.29 shows the basic concept in diagrammatic form. Some footprint calculations also allocate an additional arbitrary area (usually 10 or 12 per cent of the estimated total footprint) for 'biodiversity land'. This is land that is used as nature reserves or is otherwise managed to provide habitat for additional biodiversity.

The results of footprint calculations are possibly instructive. London has a calculated footprint of 32 million hectares, although its actual area is only 0.16 Mha. If we consider all the other inhabitants of the UK, their average per capita footprint was estimated as some 5.2 ha (The Earth Council, 2002). Note that this does not relate to any particular area of land, but is calculated on the basis of the average rates of production of food/timber or storage of carbon across the whole world.

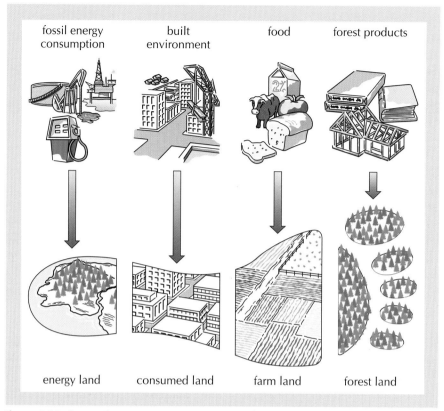

Figure 4.29 Converting consumption into equivalent areas of land.
Source: After Phil Testemale in Wackernagel and Rees, 1996, p.67.

○   If the population of the UK is 58.6 million, how does the ecological footprint of the whole population compare to the total UK area of 24 Mha?

●   The total footprint is 58.6 million x 5.2 ha, which is nearly 13 times the actual total area.

These calculations suggest that the UK needs many times its actual area to be sustainable. The situation is not quite as bad as it seems, since average productivity of land in the UK is well above the world average. Allowing for this, the UK has effectively 1.7 ha of 'average productivity land' per person, a deficit of 3.5 ha per person. Countries such as the USA and The Netherlands also have deficits of around 3.5 ha per person.

It has been estimated that 10 out of 52 countries for which calculations are available have footprints that are less than their effective productive area, so countries like the USA can 'import' area to sustain themselves. Unfortunately, one analysis (The Earth Council, 2002) suggested that even adding in the accessible and productive areas of the oceans, there is still only 2.1 ha of productive area per person, and allowing 12 per cent of this for 'biodiversity' reduces the available land to 1.7 ha per capita. The global average per capita footprint is 2.8 ha.

○   How many 'Earths' do these figures suggest we need to support current populations? And how many would we need if the current population all had a footprint equal to that of the UK?

●   With the current average footprint, we would need something like 2.8/2.1 or 2.8/1.7, giving 1.3 to 1.6. If the average footprint were to rise to the present UK level, we would need 5.2/1.7 = 3.1 'Earths'. Unfortunately, we have only one!

So how does humanity currently survive, or are the calculations nonsense? For a start, we do not yet attempt to plant enough forest to absorb the carbon dioxide released by fossil fuel burning. A second doubt surrounds the concept of productive land, and its average global productivity. There is considerable uncertainty both about the actual areas of land of different types, and of the typical or potential rate of production of relevant materials that each could sustain. Potential production is also affected by the technology that is used. Technology that improves production per hectare would reduce the footprint of any chosen subject.

There is thus a whole series of questions raised by footprinting and in 2002 it was a major source of debate. A number of websites and journal articles publicized footprint calculations and called for action to reduce footprints – or decried the concept, highlighting its flaws. This activity provided an interesting example of the way in which environmental information is created and disseminated **(Smith, 2003)**. The web has also provided a means whereby individuals can calculate their own household footprints. You may like to seek out such sites for yourself, and consider what might be done in the light of such calculations.

Summary

* A calculation of the land area needed to sustain some human grouping can be used as an indicator of the impact of that grouping on its environment.
* The ecological footprint calculation sums the area actually occupied by buildings and so on, with the calculated area of averagely productive land needed to supply food and timber and to absorb any emissions of carbon dioxide from fossil fuel use, plus possibly an allowance for biodiversity.
* Calculated on this basis, most developed countries have a footprint larger than their actual surface area of productive land.

Activity 4.10

Outline the major aspects of your own household's 'ecological footprint' compared to the average household in your area.

For a comparison with my evaluation, see the answer at the end of the chapter.

8   Conclusion

In a sense, land is the most important realization of the concept of space as it is used in this series of four books. Humans are completely dependent on land for survival, and attempting to gain access to or control over land has been a dominant cause of warfare throughout human history. From the preceding sections, you should now be able to describe the way in which land is used and modified by humans, to provide for what we see as our needs and wants. You should recognize that human activities are affected by the differing character-istics of land across space, and they can also change that land over time. The effects of human activity and of natural processes are not always restricted to the specific scale on which these processes take place. For example, the interaction of plant nutrients with the microscopic particles in the soil affects the nature of nutrient cycles, with results across a much wider area. You should be aware of the need to consider, too, the time dimension of change in land, as for example in the way in which humans react to change brought about by agricultural activity. You should recognize that there is great uncertainty over the nature of land and associated resources on a global scale, because of the difficulties in classifying and measuring environments, and because of the interaction between resources, economics and technology.

The fundamental importance of land is reflected in the idea of the ecological footprint as a convenient overall indicator of human impact on environments. Footprint calculations suggest that human activity at current levels does not have sufficient productive land available globally to be sustainable. The concept of the footprint emphasizes the spatial interdependence of human activity across the world. You should now be able to consider your own activities and lifestyle, and relate them to their environmental impact, as indicated by the ecological footprint. Potentially, this should enable you to consider what action you could take to change your own environmental impact. This is a fundamental concern of **Blowers and Hinchliffe (2003)**, the fourth book in this series.

References

Bingham, N. (2003) 'Food fights: on power, contest and GM' in Bingham, N. et al. (eds).

Bingham, N., Blowers, A.T. and Belshaw, C.D. (eds) (2003) *Contested Environments*, Chichester, John Wiley & Sons/The Open University (Book 3 in this series).

Blowers, A.T. and Elliott, D.A. (2003) 'Power in the land: conflicts over energy and and the environment' in Bingham, N. et al. (eds).

Blowers, A.T. and Hinchliffe, S.J. (eds) (2003) *Environmental Responses*, Chichester, John Wiley & Sons/The Open University (Book 4 in this series).

Blowers, A.T. and Smith, S.G. (2003) 'Introducing environmental issues: the environment of an estuary' in Hinchliffe, S.J. et al. (eds).

Carson, R. (1962) *Silent Spring*, London, Hamish Hamilton.

Earth Council, The (2002) www.ecocouncil.ac.cr, website of The Earth Council [accessed October 2002].

Flannery, T.F. (1994) *The Future Eaters*, Chatswood, New South Wales, Reed.

Freeland, J.R. (2003) 'Are too many species going extinct? Environmental change in time and space' in Hinchliffe, S.J. et al. (eds).

Freeman, P. H. and Fox, R. (1994) *Satellite Mapping of Tropical Forest Cover and Deforestation: A Review with Recommendations for USAID*, Arlington, VA, Environment and Natural Resources Information Center, DATEX.

Fuller, R.J., Gregory, R.D., Gibbons, D.W., Marchant, J.H., Wilson, J.D., Baillie, S.R. and Carter, N. (1995) 'Population declines and range contractions among lowland farmland birds in Britain', *Conservation Biology*, vol.9, pp.1425–41.

Gomez-Pompa, A., Vasquez-Yanes, C. and Guevara, S. (1972) 'The tropical rain forest: a non-renewable resource', *Science*, vol.117 (4051), pp.762–5.

Hinchliffe, S.J. and Belshaw, C.D. (2003) 'Who cares? Values, power and action in environmental contests' in Hinchliffe, S.J. et al. (eds).

Hinchliffe, S.J., Blowers, A.T. and Freeland, J.R. (eds) (2003) *Understanding Environmental Issues*, Chichester, John Wiley & Sons/The Open University (Book 1 in this series).

HMSO (1994) *Survey of Land for Mineral Workings in England*, London, HMSO.

Hodges, C.A. (1995) 'Mineral resources, environmental issues and land use', *Science*, vol.268, pp.1305–11.

Jenkinson, D.S. (1990) 'The turnover of organic carbon and nitrogen in soil', *Philosophical Transactions of the Royal Society, Series B, Biological Sciences*, pp.53–60.

Leach, G. (1976) *Energy and Food Production*, Guildford, IPC Science and Technology Press.

Pretty, J.N. and Howes, R. (1993) 'Sustainable agriculture in Britain: recent achievements and new policy challenges', *IIED Research Series*, vol.2, no.1.

Sedjo, R.A. and Clawson, M. (1983) 'How serious is tropical deforestation?', *Journal of Forestry*, 81, pp.792–93.

Smith, J.H. (2003) 'Making environment news' in Bingham, N. et al. (eds).

Wackernagel, M. and Rees, W. (1996) *Our Ecological Footprint*, Gabriola Island, British Columbia, New Society Publishers.

Answers to Activities

Activity 4.2

(a) Totalling the first two uses gives about 36.6 per cent in agricultural use.

(b) It's not possible to calculate a figure for natural vegetation, since this represents an unknown proportion of both the 'Other land' and 'Forest and woodland' categories. Even some of the 'Permanent pasture' land might be 'natural'.

Activity 4.3

The main forested areas were South America, North and Central America and the former USSR, areas where forest-type biomes are prevalent. Over the period 1961–1981, there were declines in forested areas, particularly in North and Central America, Asia and in South America, although there appeared to have been some recovery by 1991. The forested area of the EU increased consistently over the whole period. Overall, there seemed to have been little change in the total area classified as forested, but then the data for 2000 suggest that these earlier figures are dubious, and that there had been significant reductions in forested area. One commentator even suggested to me that the earlier figures should be ignored completely!

Activity 4.4

The answer has to be yes, in this case, since the organic matter levels in the continuously cropped soil with fertilizers remains higher than that without fertilizer. Even without fertilizer, levels did not fall greatly and the use of farmyard manure raised them well above the starting level. This may suggest that the organic matter level in pre-agricultural use would be even higher, so that just the change to cropping had resulted in some loss.

Activity 4.5

(a) Over the whole period, South America shows the greatest increase, from an index value of 40 in 1961 to over 140 in 2001.

(b) The only per capita data refer to the whole world, and this shows a general increase. This global figure conceals wide variation between regions, and there appeared to have been little increase from 1985 to 1993 and, arguably, over the last few years shown.

(c) European production grew steadily until about 1985, after which it remained pretty much constant. There are two possible explanations for this:
  • production reached some physical or biological limit
  • production was deliberately restricted to reduce surpluses in the European Union.

The latter is the more plausible.

Activity 4.6

(a) The seriousness presumably depends on the total amount of energy involved, which depends both on the energy need per kg of fertilizer produced, and the amount of fertilizer needed. The nitrogen as nitrate and nitrite anions in soil is very susceptible to leaching, so it is likely that more nitrogen fertilizer will be needed than phosphorus or potassium. This exacerbates the problem caused by the high energy demand per kilogram.

(b) World production is 80 M tonnes, or 80×10^9 kg. Assuming each kg requires 80 MJ for synthesis, this represents a total of $80 \times 80 \times 10^{15}$ J, or 6.4×10^{18} J. Total energy use by humans across the world is approximately 3×10^{20} J.

Activity 4.7

There are several major effects. Firstly, agricultural land is usually obtained by clearing forests, prairies or savannah that would otherwise represent the prevailing biome in many areas. Introducing agricultural livestock can have a major effect on sensitive faunas and floras as illustrated by what has occurred in Australia and the Americas. Having made these major changes to communities, recent developments in agriculture have had further direct and indirect effects on other species through the use of pesticides. The use of fertilizers involves substantial use of fossil fuel for their production, and excess use can result in pollution of water bodies. Use of water for irrigation can change the nature of soil.

Since agriculture is a major user of land, such effects are clearly important, but how we value them depends on how hungry we are!

Activity 4.8

In all cases, the reserves have increased over the period concerned, except for lead which possibly peaked in the 1970s. This may represent the same general effect, since demand for lead was highest then and has since declined with the phasing out of lead as an additive in petrol. This would have the effect of reducing the price of lead and, as a result, make the lowest concentrations of lead ore uneconomic to mine, thus decreasing the estimated reserve.

Activity 4.9

The area occupied by mineral extraction is small, and concentrated in a few places, so impact is likely to be uneven, but with local damage possibly severe. The small number of discrete sites are relatively easy to control, so limiting damage and ensuring remediation should be feasible. Agriculture, in contrast, is ubiquitous, and so is much more difficult to regulate. The effects are likely also to be uneven, but probably less so than for minerals. It is a matter of debate as to which has the greater impact where it occurs.

Activity 4.10

The components of the footprint are land for buildings, food, forest products and 'energy land'. Here's my evaluation of the footprint of my own household.

My household occupies a larger than necessary property, but with only two of us, our need for food consumption land is probably less than the average household. Our annual consumption of manufactured forest products as paper is very high, and we do use wood as a fuel, in addition to oil for heating. Our fossil fuel consumption relates not just to the heating and car use, but we also have to take a share of the fuel used in growing, processing and transporting our food and other consumables. Because of my particular background, we try to avoid foodstuffs that have involved heavy energy use for transport (fresh salads from Kenya, for example) or, to a lesser extent, for storage (freezing is particularly energy-intensive). As a household, we try to limit car use, but living in a rural area, other forms of transport are rarely an option. Buses and trains use less energy per person kilometre, and hence have a lower energy land footprint than private cars.

Chapter Five

Water

Mark Brandon and Sandy Smith

Contents

1   Introduction: water, water everywhere

Figure 5.1 A view of the Earth from Apollo 17 in 1972. The blue areas are oceans, the white swirls are clouds, the uniformly white area is the Antarctic ice cap and the brown area is Africa.

When astronauts first went into space, they were captivated by a vision of the Earth that had never been seen before – an Earth predominantly blue in colour (Figure 5.1). After photographs such as this were published, the Earth became known as 'the blue planet'. The Earth looked blue because the oceans cover 71 per cent of its surface. Water in other forms is also visible in Figure 5.1: as water droplets in clouds, and ice in the polar ice caps.

Water is essential for life – all living organisms contain water. This may not be initially obvious, but water is always a major component. In fact most living organisms consist of over 60 per cent water (Table 5.1).

Table 5.1   Water content of various plants and animals

Plant	Water content/ % total mass	Animal	Water content/ % total mass
lettuce	96	codfish	82
tomato	93	earthworm	80
strawberry	89	frog	78
parsnip	83	chicken	74
banana	71	herring	67
apple	65	human	65
peanut	5	dog	63

Figure 5.2 Tourists on the Galapagos Islands with the Galapagos giant tortoises.

There are many places in which the supply of fresh water is limited and choices have to be made, for example, on remote islands such as the Galapagos in the Pacific Ocean, where the only fresh water comes from the limited rainfall. Water demand on the Galapagos is increasing with increasing tourism (Figure 5.2), and further development is limited (at least in part) by shortage of fresh water. But the Galapagos Islands are surrounded by water – the Pacific Ocean – isn't that the same stuff? Only up to a point: seawater contains around 3.5 per cent salts and, although that's not much, it's sufficient to make you ill if you drink it. The difficulty was summed up by Coleridge's 'Ancient Mariner', on board a ship becalmed in the Pacific Ocean beneath a blazing Sun (see Figure 5.3):

> 'Water, water, every where, And all the boards did shrink.
> Water, water, every where, Nor any drop to drink.'

(Samuel Taylor Coleridge, *The Rime of the Ancient Mariner*, 1797–1798)

Figure 5.3 *The Rime of the Ancient Mariner*, illustration by Gustave Doré.

Sea water can be used by humans for some purposes such as flushing WCs. Salt can also be removed by a process called **desalination**. The cheapest way to do this is to use the natural energy of the Sun to evaporate water from sea water (this leaves the salt behind, as we saw in the previous chapter). Faster desalination methods use a different energy source, oil, for example, and so can be expensive. However, whereas in Kuwait oil is cheaper than water and so is used to power desalination plants, many places, like the Galapagos, do not have a cheap supply of energy and this is not economic.

desalination

The ocean is still useful to humans on the Galapagos. It is used for fishing and transport of goods and people, as well as being useful to other organisms as an environment for marine life. These human uses can cause problems: when the oil tanker *Jessica* was wrecked on the Galapagos in 2001 it caused oil pollution to the seas and beaches, and damage to the unique life of the islands (Figure 5.4).

In this chapter we will investigate the key questions of how and why the fresh water and ocean environments are changing. We will consider both natural changes and changes caused by humans (such as the oil spill from the *Jessica*), their variations in space and time, and the

Figure 5.4 The wreck of the oil tanker *Jessica* off the coast of the Galapagos Islands, January 2001.

consequences and significance of the changes. The next section will investigate how water moves around the Earth in the hydrological cycle. Fresh water resources will be examined in Section 3 and, finally, Section 4 will concentrate upon water in the oceans.

2   The hydrological cycle

The concept of the hydrological cycle is introduced in **Blowers and Smith, 2003.**

Water molecules do not stay in one place on the Earth, they move between the land, ocean, ice caps and the atmosphere in the hydrological cycle (see Figure 5.5).

A change in the amount of water available, either over space or time, can be caused by changes in any component of the hydrological cycle. In this section we will study each component of the cycle, to see how and why it can produce these changes.

2.1 Precipitation

We see water moving through the hydrological cycle when it rains. This process starts with atmospheric water vapour cooling (usually when air rises, for example by being blown over a mountain). Some water vapour condenses into droplets, which form clouds (just like the condensation that forms when you breathe out on a cold day). Once they have condensed, the small droplets join together and eventually grow too large to stay in the atmosphere, so they begin to fall. If the air temperature is below the freezing point of water, they freeze and become hailstones. Snow is formed if the air temperature is below freezing at the point at which the water vapour condenses. Here the vapour freezes directly to ice. The general term for rain, hail and snow is **precipitation.**

precipitation

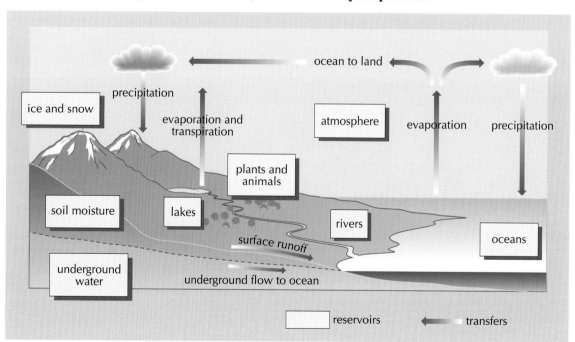

Figure 5.5 The hydrological cycle moves water around the Earth, driven by gravity and the Sun's energy.

Precipitation is measured using a rain gauge. This is a container that collects and records the amount of precipitation in terms of the depth of water in the container. Precipitation is therefore measured as a *depth* of water, usually in units of millimetres. You can easily create a simple rain gauge from a straight-sided container, such as a tin can. Leave it outside during a rainstorm and, when the rain has finished, measure the depth of water to get the value for the precipitation during the rainstorm. More sophisticated rain gauges have designs that prevent them being blown over, or knocked over by animals; they also minimize evaporation and can record the depth of water automatically. Figure 5.6 shows the average annual precipitation around the UK.

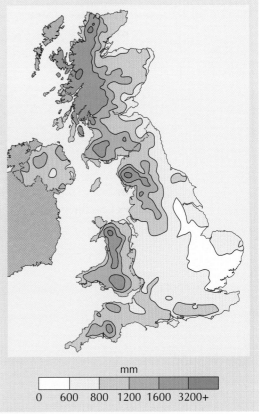

mm

| 0 | 600 | 800 | 1200 | 1600 | 3200+ |

Figure 5.6 Average annual precipitation in the UK, 1961–1990, measured as a depth of water. *Source*: adapted from the UK Meteorological Office's website.

Activity 5.1

(a) What are the average annual precipitation values in London, Cardiff, Edinburgh and Belfast?

(b) What is the range (minimum and maximum values) of average annual precipitation in the UK?

For the answers, go to the end of this chapter.

Regional variation in precipitation is caused by differences in altitude and in the paths of weather systems (see Chapter Six) across the UK.

The mean annual precipitation on Earth is around 1,000 mm. However, there are large variations about this average in different places. Some areas get less than 250 mm annually; these are usually deserts, such as in the Middle East, North Africa, north-central Asia and central Australia (see Figure 5.7). The annual precipitation in other areas can reach as much as 12,000 mm (for example the Amazon Basin and parts of South and South-East Asia). Precipitation may also vary significantly with time, as well as spatially. In some regions precipitation is seasonal, as on the Indian subcontinent where the south-west monsoon brings rain for a few summer months (Figure 5.8).

Activity 5.2

The average annual precipitation for Cambridge and New Delhi is very similar, although markedly different in seasonal distribution. By how many millimetres does the monthly precipitation vary in (a) Cambridge and (b) New Delhi?

For the answers, go to the end of this chapter.

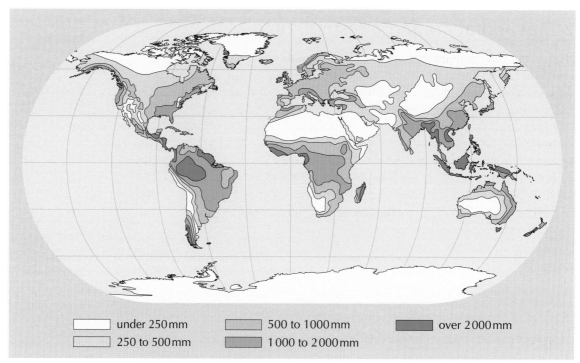

Figure 5.7 Average annual precipitation around the Earth.

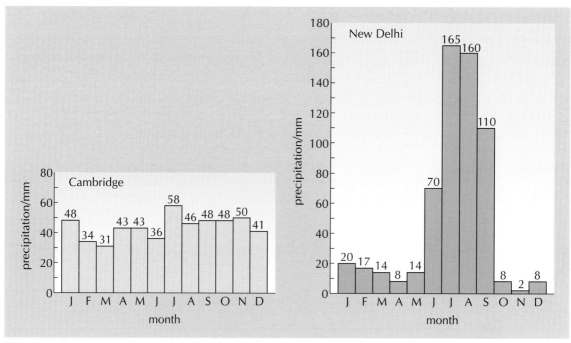

Figure 5.8 Average monthly precipitation for Cambridge, England, and New Delhi, India.

2.2 Evaporation and transpiration

How much water is in the atmosphere? If all the atmospheric water were to fall evenly over the Earth, it would form a layer only about 30 mm thick. The atmosphere can only keep supplying rain because water is cycled from the Earth's surface to the atmosphere by evaporation and transpiration.

Evaporation depends on temperature (evaporation increases as temperature increases), humidity (how much water the air contains relative to the maximum amount of water vapour it can hold) and wind speed. The greater the humidity, the less the evaporation. Wind carries water-saturated air away from the surface and allows more water to evaporate, so a higher wind speed increases evaporation.

evaporation

○   Which areas will have the highest potential for evaporation?

●   Areas with high temperature, low humidity and high wind speed, such as low-latitude deserts (close to the Equator).

Activity 5.3

Figure 5.9 shows the average monthly variation of evaporation 1956–1962 from a lake in the Thames Valley, England. How does the evaporation change over the year, and why do you think it does this?

For the answer, go to the end of this chapter.

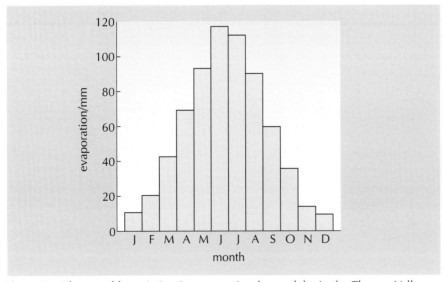

Figure 5.9 The monthly variation in evaporation from a lake in the Thames Valley, England.

The other process that transfers water from land to the atmosphere is **transpiration** by plants. They draw water from the soil through their roots, *transpiration* transport it to the leaves from where it evaporates into the atmosphere. Transpiration can transfer significant amounts of water. For example, a large oak tree transpires around 400 litres a day in the summer.

2.3 Surface water

The flow of rivers varies with time, usually seasonally, in a way that depends on the climatic zone that the river flows through. A graph of the discharge of a river over time is called a **hydrograph** (see Figure 5.10).

hydrograph

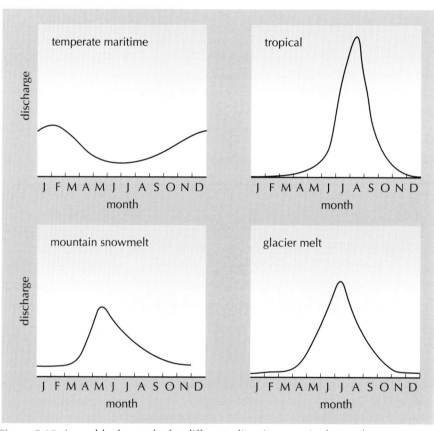

Figure 5.10 Annual hydrographs for different climatic zones in the northern hemisphere. The first two hydrographs are mainly rainfall dependent and the last two are mainly temperature dependent.

Activity 5.4

The UK is in a temperate maritime climatic zone. Explain, using your knowledge of precipitation and evaporation in the UK, the shape of the temperate maritime hydrograph in Figure 5.10. Refer back to Figures 5.8 and 5.9 for clues to the answer.

For the answer, go to the end of this chapter.

The variation in seasonal precipitation is the main reason for the shape of the tropical hydrograph in Figure 5.10. Within the tropics and the equatorial zone, there are usually wet and dry seasons, or even two wet seasons, which control the natural flow of the regions' rivers. The other two hydrographs in Figure 5.10 are controlled by temperature. When the river catchment (the area contributing to the flow of the river) includes mountain snow or glaciers, melting of the snow or ice takes place in late spring or summer, giving a peak flow around that time.

2.4 Underground water

There is a substantial store of water under the Earth's surface. The most obvious underground water is seen in vast limestone caverns, such as those in the Derbyshire Peak District. Here water flows from one cavern to another along underground rivers. Although underground caverns are fairly rare, water does commonly exist underground, *within* soil and rocks. You read in Chapter Four that soils can contain small pores; rocks can also have small pores and these pores can hold air, water or even oil or gas (see Figure 5.11).

Figure 5.11 An electron micrograph of a sandstone rock. Pores make up 31 per cent of this rock's volume.

In soil or rock near the ground surface, pores are usually partially air-filled, but at greater depths, below a level called the **water table**, the pores are filled with water. Water below the water table is called **groundwater** (Figure 5.12).

water table

groundwater

A rock that is sufficiently porous to store water and also allows water to flow through it, is called an **aquifer**. These can act as underground reservoirs, into

aquifer

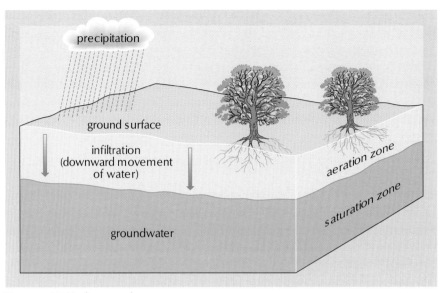

Figure 5.12 Underground water.

which wells can be sunk and water pumped out when it is needed. Only some rocks are good aquifers; in England the chalk rock in the south-east is notably good.

2.5 Reservoirs, transfers and residence times

reservoirs

The places where water is stored in the hydrological cycle, on and under the land, in the oceans and in the atmosphere, are called **reservoirs** (not to be confused with human-built reservoirs). Figure 5.13 shows the amounts of water stored in each of these reservoirs. In terms of size, the ocean is the largest by far, with 96 per cent of the total. The amount of water in the atmosphere is relatively small. On land the largest reservoir is ice and snow, with comparatively little surface water in lakes and rivers. What may be surprising is the relatively large amount of water underground. Figure 5.13 also shows the rates of transfer between the reservoirs, and can be used to investigate the cycle in more detail and explore the

renewability

renewability (the rate at which a resource is naturally replenished) of water in each reservoir.

○ Is the rate of transfer for water precipitated onto land and the rate for water evaporated and transpired from land the same?

● No, water is precipitated onto land at a faster rate $(107 \times 10^{15}\,\text{kg y}^{-1})$ than evaporation and transpiration from it $(71 \times 10^{15}\,\text{kg y}^{-1})$.

As Figure 5.13 demonstrates, the rate of precipitation onto land exceeds the rates of evaporation and transpiration, resulting in an excess. This excess of precipitation over evaporation and transpiration must eventually return to the atmosphere, otherwise the land reservoirs would increase in size and the cycle

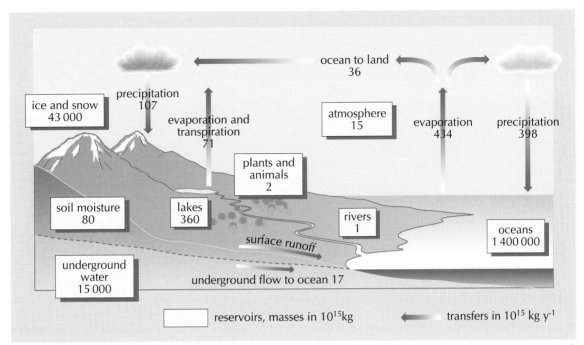

Figure 5.13 Reservoirs and transfer rates of the hydrological cycle. Boxes represent reservoirs where water is stored, with amounts of water in units of mass (10^{15}kg; this means that the number on the figure is multiplied by 10^{15}kg to get the value, for example there is 15×10^{15}kg of water in the atmosphere reservoir). Blue arrows show transfers of water between reservoirs, with the annual transfer rate, in units of mass per year (10^{15}kg y^{-1}).

would not be in balance. Figure 5.13 shows that there are two other ways of transferring water from the land: by surface runoff and underground flow (this way water reaches the oceans). By both of these ways, water ultimately returns to the atmosphere.

○   Are the rates of transfer of water to and from the atmosphere in balance?

●   Yes: transfer to the atmosphere is by evaporation from the oceans (434×10^{15}kg y^{-1}), and from the land (71×10^{15}kg y^{-1}); and transfer from the atmosphere is by precipitation to the oceans (398×10^{15}kg y^{-1}) and land (107×10^{15}kg y^{-1}). These balance, as $434 + 71 = 505$, and $398 + 107 = 505$.

The renewability of water in any reservoir depends on how quickly that reservoir is replenished. What is the renewability of water in the atmosphere? We investigate this by comparing the amount in the reservoir with the rate of input or output to calculate a value called the **residence time**. The rate of input must be   residence time the same as the rate of output for any reservoir, otherwise the reservoir would change in size and the cycle would not be stable.

residence time = amount in reservoir/rate of transfer from or to the reservoir

For example, for the atmosphere:

$$\text{residence time} = 15 \times 10^{15} \, \text{kg} / 505 \times 10^{15} \, \text{kg y}^{-1}$$

$$= 0.0713 \text{ years (or about 11 days)}$$

Residence time (Table 5.2) is helpful for considering the usefulness of any component of the cycle for water resources. The atmosphere stores very little water (0.001 per cent of the total), but has a residence time of only about 11 days, so water in it is generally renewed relatively rapidly. Residence time is very variable, depending on which of the reservoirs is considered, from days to 10,000 years.

Table 5.2   Residence times for water in the hydrological cycle

Reservoir	Mass of water/10^{15} kg	Percentage of total water	Residence time
ocean	1,400,000	96	about 3,000 years
ice and snow	43,000	2.9	about 10,000 years
underground water	15,000	1.0	a few weeks to 10,0000 years
lakes	360	0.025	about 10 years
soil moisture	80	0.005	a few weeks to 1 year
atmosphere	15	0.001	about 11 days
rivers	1	0.00007	a few weeks
plants and animals	2	0.00014	a few days to several months

Note: the figures in this table do not total 100 per cent due to rounding up.

Summary

Throughout this section, we have focused on the key concepts of space and time as applied to water. Spatially, most of the water is in the oceans. On land, most of the water is in the form of ice and snow or underground. Surface water on land has a variable distribution, depending on precipitation, evaporation and transpiration. There is also variation of water with time, such as seasonal rainfall in monsoons and the seasonal variation of river flow. The hydrological cycle links water in the different reservoirs of the cycle. The rates of movement between reservoirs affect the renewability of water and this is indicated by the residence time.

3   Fresh water resources

In Section 3 we will consider the key questions of how and why fresh water environments change, and how we determine the consequences and significance of changing water environments. Most, but not all, of these changes are caused by human usage of fresh water.

3.1 Water for life

Plants and animals use water in many ways: in **metabolism**; as a living environment (for example whales in oceans, fish in rivers, algae in ponds); or it may drive the pattern of their lives (for example animal migrations in response to seasonal rain).

Metabolism is the sum of all the chemical and physical processes within a living organism.

A basic amount of water is needed by humans each day for survival. In the UK this is about 2.5 litres per person, part supplied by drinking, part from water in food. A person doing a lot of physical work, or in a hotter climate, would need more, up to 12 litres. Without adequate water intake dehydration occurs rapidly. With loss of water equal to 5 per cent of body mass (around 3 or 4 kg), you would feel incredibly thirsty, and a 10 per cent loss would make you seriously ill.

Humans also depend on water for more than just drinking. Figure 5.14 illustrates how a household uses water. We use far more water when it is freely available from a tap than when it must be carried from a river or a well to a house.

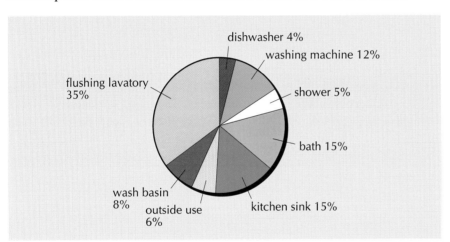

Figure 5.14 The average household fresh water consumption for England and Wales, by purpose, for 1996.

Life in the developed world has come to depend on unlimited supplies of water being available at the turn of a tap. In the UK, the average domestic water use is 136 litres per person per day. Water in the UK is also used for commercial and industrial purposes, electricity generation and agriculture, at an average of 604 litres per person per day (Figure 5.15a). The average domestic water use in developing countries is estimated at just 10 litres per person per day.

On a global scale, the greatest overall use of fresh water is for irrigation, 69 per cent (Figure 5.15b) although this varies spatially. For example, in the UK irrigation is only a few per cent of total fresh water use, in Asia it is 86 per cent. However, water is a resource in which 'used' is a relative term since it is returned to the hydrological cycle after use, but usually with a change of quality.

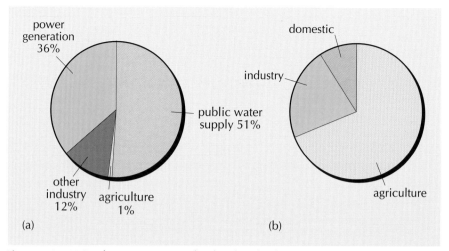

Figure 5.15 (a) Fresh water use in England and Wales for 1996. (This excludes seawater used for cooling purposes by some power generation plants.) The public water supply is used by offices, some industries and farms as well as by domestic users. (b) Global fresh water use, by purpose.

3.2 Water supply

Water in rivers and lakes is the most accessible part of the hydrological cycle for human and animal use. However, over a half of the Earth's lake water is saline and so is of little use for water supply (for example the Dead Sea, see Figure 5.16; although called a 'sea' it is technically a lake). About 80 per cent of the fresh water in lakes is in only a few large lakes, including the Great Lakes of North America and Lake Baikal in Siberia. This leaves much less lake water available on a more local scale for water supply.

Activity 5.5

Compare the relative suitability of rivers and lakes for water supply in terms of quantity and renewability, using Table 5.2.

For the answer, go to the end of this chapter.

Figure 5.16 The Dead Sea, in the Middle East. This inland lake contains large quantities of water, but it is even more saline than sea water, so can't be used for a water supply.

Artificial reservoirs have long been used to store surplus river water. The earliest known reservoirs are 5,000 years old and in Arabia. Here farmers used craters of extinct volcanoes as storage tanks. Reservoirs can also generate hydroelectric power and protect against floods by trapping and storing floodwater. For example, the Aswan High Dam on the River Nile in Egypt generates electricity, reduces seasonal flooding, and also provides water for irrigation. Reservoirs on the Colorado River are similarly multi-purpose (Box 5.1).

Box 5.1 The Colorado River

The Colorado is a large river in the USA with vast seasonal fluctuations in flow. In 1920, its flow varied from 50 – 2,600 m^3s^{-1} (cubic metres per second), see Figure 5.17. This seasonal variation is caused by mountain snowmelt in the upper parts of the catchment. Further downstream, the river flows through deserts in Nevada and Arizona and so its water is very useful. As most of this flow used to occur in seasonal periods, it was of limited use for irrigation. The snowmelt also produced annual flooding.

To regulate the river flow for water supply, generate hydroelectric power and prevent flooding, the river has been dammed to create reservoirs. The first major dam, the Hoover Dam, was completed in 1935, forming the reservoir of Lake Mead. This holds most of the spring snowmelt and prevents very high flows and most flooding (Figure 5.17).

○   How has the creation of Lake Mead affected the discharge of the Colorado below the Hoover Dam?

●   The discharge is more uniform – there are fewer very large peaks.

Even after the completion of Lake Mead there were times when floodwaters filled the lake and had to be released downstream, so in 1963 another reservoir (Lake Powell) was created further upstream to trap this excess floodwater.

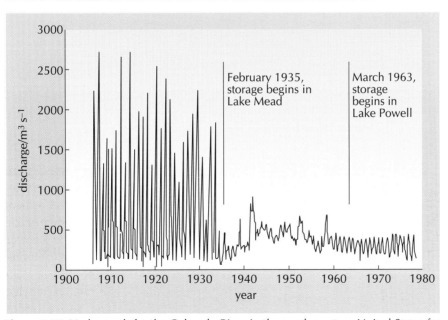

Figure 5.17 Hydrograph for the Colorado River in the south-western United States for 1906–1979, measured downstream of the Hoover Dam. Lake Mead is the reservoir behind the Hoover Dam.

Reservoirs, although very useful, are sometimes causes of major environmental changes:

- Loss of land
  Reservoirs may flood large areas of land. Lake Nasser, the reservoir created by the Aswan Dam, has an area of 6,000 km^2. Its planning provoked much opposition, particularly for the destruction of important artefacts, such as the potential drowning of the ancient temples of Abu Simbel. However, the dam was constructed and some of the temples were moved (see Figure 5.18). In the UK, also, some 250 km^2 of land are covered by reservoirs, some of which have drowned whole villages, and displaced people to other areas. The need for water (often to supply a large town) has been thought to be more important than the small rural village. The largest reservoir in the UK is Kielder Water in Northumberland, which covers about 10.5 km^2. Drowning land under a reservoir also causes the loss of land as a resource.

Figure 5.18 The Abu Simbel temples in Egypt, during their reconstruction in 1966.

- Ecological changes
  The creation of a reservoir not only changes land to water, but also produces changes both upstream and downstream of the reservoir. The gradient of a river upstream of a reservoir may be reduced. This slows the river and changes its character, causing deposition of sediment and changes to the natural vegetation and animal life of the river. Downstream the flow will reduce, also affecting plant and animal life. Migrating fish species, such as salmon, may also be prevented from reaching their spawning grounds upstream of a reservoir.

- Dam failure
  Dams may collapse, releasing large amounts of water downstream, killing people and animals as well as destroying buildings. Collapse may be caused by inappropriate construction, failure of the underlying sediments or rock, overfilling or earthquakes.
- Sediment loss to agriculture
  The River Nile used to flood its banks once a year, and the sediment in the river was deposited on the surrounding land, renewing the nutrient supply. This floodwater is now trapped behind the Aswan Dam, the sediment has settled out in the reservoir, and so synthesized fertilizers have to be used in the Nile Valley.
- Erosion
  The Nile Delta on the Mediterranean coast has formed over the years from sediment carried down by the river. Since the creation of the Aswan Dam, little new major sediment reaches the delta and waves are causing erosion, destroying the delta habitat for marine organisms and its important fishing grounds for humans.
- Soil salinization
  The change from annual flooding by a river to irrigated agriculture can cause salinization. Salinization causes a decline in crop yields until eventually the soil may become useless for agriculture, as was discussed in Chapter Four.

Groundwater is often also useful as a water supply. In England, for example, about a third of the water abstracted is groundwater, but in Scotland and Northern Ireland the underlying rocks are older, and usually have less pore space so they make poorer aquifers. These latter regions only get about 5 per cent of their water from groundwater.

Activity 5.6

What would be the environmental advantages and disadvantages of using groundwater instead of river water as a water supply?

For the answer, go to the end of this chapter.

An aquifer may cause the formation of an **oasis** where groundwater that fell as rain in mountains (a distance from the oasis) flows underground through an aquifer to emerge at the surface in an area with less precipitation, typically a desert (Figure 5.19). Water flow through aquifers can be very slow, and the water from such an aquifer may have flowed a long way from where it fell, taking a long time to reach the point of extraction. In some parts of northern Africa and the Middle East, groundwater is being extracted that fell over 10,000 years ago, a time when the climate was wetter during the last interglacial Stage. These aquifers contain water that is not being currently renewed at the same rate at which it is being used, so it is a **non-renewable resource** and is being used **unsustainably**.

non-renewable resource

unsustainable

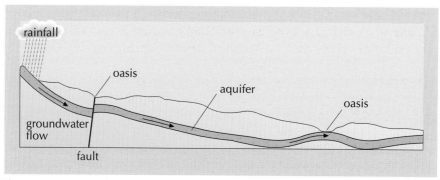

Figure 5.19 An aquifer can give rise to oases in desert regions, either by seepage up a fault through the rock, or by exposure of the aquifer at the surface.

Although the underground water reservoir of the hydrological cycle is much larger than the lake and river reservoirs (Table 5.2), unfortunately, the deeper groundwater is too saline to use as a resource. We have seen that some of the lakes are also saline. The other main fresh water reservoir in the hydrological cycle, ice and snow, is generally too remote to be useful as a water supply. So only a very small part of the water in the hydrological cycle (in rivers, some lakes and shallower groundwater, about 1 per cent of the total) can be used for water resources.

Activity 5.7

Discuss how the difference in the mean global annual precipitation (around 1,000 mm) and the range of monthly precipitation between Cambridge and New Delhi (Figure 5.8) would affect water resource planning both in Cambridge and in New Delhi.

For the answer, go to the end of this chapter.

3.3 The purity of water

The water environment can also change in quality as well as in quantity. No natural water found on Earth is 'pure'; every sample contains more than just H_2O. Not only solid substances dissolve in water, gases do also. For example, rainwater contains dissolved gases from the atmosphere. One of the most important of these gases is oxygen, which is also found dissolved in river water and sea water. The dissolved oxygen is extracted from water by many aquatic organisms, such as fish. Water without oxygen, or with low levels of dissolved oxygen, cannot support a wide range of aquatic life. Rainwater that has a high percentage of the dissolved gases of sulphur and nitrogen oxides is acidic; this **acid rain** is investigated in more detail in Chapter Six. Acid rain can damage forests, and also causes acidification of streams and lakes, damaging the organisms living in them.

acid rain

Ideally, all fresh water resources should be of a certain quality to be useful for their purpose, whether it is for drinking, industry or agriculture. It is not necessary or even desirable for drinking water to be absolutely pure. For example, dissolved calcium is a useful substance for healthy bones and teeth, but drinking water should not contain significant amounts of harmful substances, such as dissolved poisons, or bacteria and viruses that can cause disease.

High standards of hygiene in water supply are not always maintained, even in the developed world (which usually has a piped water supply) where the effects of high nitrate levels (mentioned in Chapter Four), sewage pollution and high salinity are often of concern. In the developing world drinking water usually comes from the nearest river or well. Contaminated water containing bacteria, parasites and viruses is a major cause of death, particularly among infants. This may seem remote from the UK, but in the nineteenth century the River Thames was so polluted by discharges of untreated sewage into it that it caused outbreaks of typhoid fever (Figure 5.20 and Box 5.2).

Box 5.2 Pollution and clean up of the River Thames

Up until the nineteenth century, it was common practice for people in London to take their buckets of sewage down to the Thames to empty, and then refill with river water for domestic use. When the population of London was small, this organic pollution was broken down by bacteria in the river water. This is a natural water cleansing process during which the bacteria use up dissolved oxygen for respiration. We now know that many bacteria and other micro-organisms in river water are harmless, and are a natural part of the aquatic food chain. Diversity of life in fresh water streams and lakes is an indicator of water quality: the more species, the better the quality.

As the population of London increased and flush sanitation was installed in houses, the quantity of sewage discharged into the river became so high that the river was unable to 'self-purify'; the bacteria could not get enough oxygen to cope with the sewage. The stench from the river became unbearable, and on occasions Parliament had to be abandoned. This forced legislation for the treatment of sewage before discharge to the river, reducing the pollutant load, so that it could again naturally self-purify.

Rapid growth of London during the first half of the twentieth century meant sewage treatment could not keep pace with the growing population. Oxygen levels plummeted once more, until the 1960s when tighter controls on sewage discharge brought a steady improvement in the condition of the river. In 1974, the first salmon was caught in the Thames for 141 years.

Sewage treatment (see Box 5.3) is fairly routine for developed nations, but in developing countries it is rare, with most urban sewage discharged to rivers, lakes or the oceans without any form of treatment. These rivers or lakes may also be used as water sources, with obvious problems for health. Development of sewage treatment and water supply systems is often the first municipal public

Figure 5.20 'Monster soup'; a cartoon, circa 1830, on the quality of the water from the River Thames being supplied as drinking water.

service for developing countries, and causes a dramatic improvement in public health, reducing inequalities at basic levels.

Nitrogen and phosphorus compounds (from domestic sewage, some industrial waste, and farms, especially from fertilizer over-application) cannot be removed through normal sewage treatment. When these plant nutrients reach rivers and lakes, the water can become nutrient-enriched (a process called **eutrophication**). This enriched environment drives excessive aquatic plant growth, causing

eutrophication

Box 5.3 Sewage treatment

Sewage treatment aims to renew the quality of water, by providing conditions for natural processes to reduce the concentration of polluting organic matter.

Primary sewage treatment is the mechanical removal of solid material, using screens for the coarse material, then sedimentation tanks, where most of the remaining particles settle to the bottom of the tanks. The liquid leaving the tanks still contains very fine solids and dissolved matter, and goes on to secondary treatment.

Secondary sewage treatment is a biological process, involving the oxidation and breakdown of organic material by micro-organisms (mainly bacteria), a process of respiration similar to that taking place naturally in rivers. The process is speeded up by increasing the amount of oxygen available, either by spraying the liquid over filters on which the micro-organisms live, providing a large surface area for oxygen uptake, or by agitating with paddles or bubbling air through the liquid.

an **algal bloom** where the water becomes so densely populated with algae that algal bloom light cannot penetrate the water to any depth. Without light, algae in the deeper water are unable to photosynthesize and die. The action of bacteria in decomposing the dead algae then reduces the oxygen level in the water, causing problems for other aquatic organisms.

3.4 Water shortages

Global fresh water use has increased rapidly in recent years at around 5 per cent per year. Most of this increase is in the developing world, and in countries with high population growth rates. Their water problems are increasing rapidly. On a global scale water is not scarce, but on a continental, national or local scale it often is and, with increasing expectations and demand, is likely to be more so in future.

Water shortages are often caused by non-sustainable use (which might, in some cases, also be inappropriate), for example in Central Asia. The Aral Sea was once the world's fourth largest lake, and used to be a very productive fishing site. In the 1980s the rivers that drained into the Aral Sea were diverted so that irrigation water could be used to grow cotton in the Central Asian desert regions. The Aral Sea is now rapidly reducing in size, leaving fishing ports high and dry, and increasing salinity in the water is killing the fish and ruining the fishing industry. In addition, the irrigated desert is becoming unproductive because of soil salinization.

Activity 5.8

Explain why the use of water for irrigation from the rivers that flow into the Aral Sea is (a) inappropriate (having undesirable side-effects), and (b) non-sustainable.

For the answer, go to the end of this chapter.

The Aral Sea example illustrates that there may be competing demands for a water resource that is insufficient to meet the needs of all.

3.5 Water resources and a changing climate

A major source of concern for fresh water resources in the future is climate change. This can affect the hydrological cycle and water resources in two main ways. An increase in temperature will increase the amount of water vapour in the atmospheric reservoir. This could change precipitation amounts and patterns. It is also predicted to increase the frequency of extreme events, such as droughts Climatic change, and its and floods (see Chapter Six). prediction, is discussed in more detail in Chapter Six.

For the UK, the annual precipitation is predicted to become more variable from year to year and between seasons, producing more droughts and floods. Precipitation is expected to increase in winter and autumn and decrease in spring

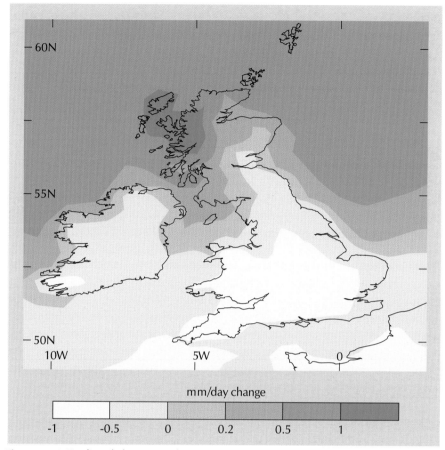

Figure 5.21 Predicted changes in the summer rainfall over the British Isles for the 2090s in comparison with a century earlier.
Source: adapted from the UKCIP's *Technical Report Number 1*, 1998.

and summer when more water resources are needed. Regional variations are also predicted (Figure 5.21).

Activity 5.9

Are the areas with predicted increases in summer rainfall in Figure 5.21 the areas where an increase would be most useful? You will also need to refer to Figure 5.6 to answer this.

For the answer, go to the end of this chapter.

Over the northern UK, precipitation intensities will increase in both winter and summer, the most intense events becoming several times more frequent than at present. This is likely to lead to greater risk of flooding. Over the southern UK, precipitation intensity, and so the frequency of flooding, will increase only in winter (see Figure 5.22). In the south, in summer, because there is a reduction in the overall amount of precipitation there are predicted to be fewer high intensity rainfall events in comparison with the present climate.

Figure 5.22 Winter flooding in York, England, November 2000.

Summary

This section has looked at changes to the water environment, caused by the exploitation of fresh water resources by humans. We interact mostly with the river part of the hydrological cycle, causing changes to the environment visible, such as the creation of reservoirs, and others that may be less visible, such as pollution. Our use of groundwater also causes environmental change when aquifer water is used unsustainably. As well as changing the quantity of water available in time and space terms, we also change the *quality* of water by pollution or eutrophication.

There are longer-term environmental changes due to human-caused climatic change. The amount of water vapour in the atmospheric reservoir is predicted to increase, changing the stability of the hydrological cycle and precipitation amounts and patterns. This could increase the frequency of droughts and floods.

4   Oceans

Table 5.2 showed that the largest component of the hydrological cycle is the oceans. Unsurprisingly this has a huge effect on the Earth. Here we consider the current state of the ocean, how it 'works', and some human uses of it with the aim of demonstrating how time and space are key tools for understanding water on Earth. We will see that some of these resources are vulnerable to changes, both natural and human-induced.

4.1 Ocean basins

Figure 5.1 showed that the Earth is clearly a blue planet. But what would we see if we could get rid of all the water? If you were in a ship, knowing the depth of water would be pretty important – if only to avoid running aground. The oceans' depths were first seriously mapped in the nineteenth century and this knowledge immediately changed the link between time and space by reducing communications time across the planet (Box 5.4). It is only during the last few decades that such maps have been of high quality and scientists have begun to understand the

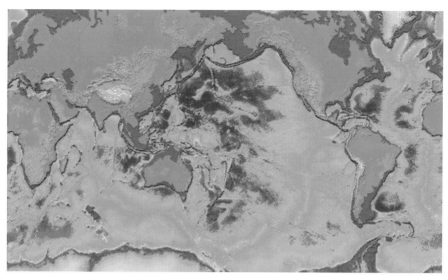

Figure 5.23 The topography of Earth. The different colours in this representation indicate different depths and heights. The deepest regions are in the oceans and are dark blue, whereas the highest regions on land are light brown.

global impact of underwater mountains and valleys on human use of the oceans, and on the Earth's climate. Figure 5.23 shows our current most accurate map of the Earth's topography.

Figure 5.23 shows a surprising amount of detail under the sea. The boundary between the different continental plates (discussed in Chapter One) can be seen down the middle of almost all the oceanic regions as light green/yellow lines. These are the mid-ocean ridges that make up the most extensive mountain range on Earth. The dark blue regions in the oceans' depths are called **abyssal plains**. They cover the greatest proportion of the Earth's surface and are the flattest regions on the planet with average gradients less than 0.05 per cent. The steepest gradients are (as the early sailors found) at the edge of the continents and are approximately 4 per cent. Within these two gradients lie most of the topography across the whole of the oceans.

abyssal plains

Gradient is the change in elevation divided by the distance over which the change occurs. This value is then multiplied by 100 to give a percentage.

Activity 5.10

How do these gradients compare with those found on land? (Think of the sorts of gradients you sometimes see on road signs.)

For the answer, go to the end of this chapter.

It is hard to appreciate the true scale of the oceans' depths. The Earth has a radius of approximately 6,370 km, and the mean ocean depth is 3.7 km. This means that, although the oceans constitute the largest reservoir in the hydrological cycle, they are no more than a thin layer on the surface of the Earth. In fact they are about the same relative thickness as the layer of skin on an apple!

We know that the shape of the oceans is changing slowly over millions of years through plate tectonics, but the time periods are so great that we consider the shape of the oceans as being static on a day-to-day basis. We shall see below that the shape of the ocean basins actually controls the ocean circulation.

Box 5.4 The telegraphic plateau across the North Atlantic

The first reason for seriously mapping the oceans' depths was commercial. The industrialized world had discovered telecommunications and speculators wanted to lay a telegraph cable across the Atlantic. But how much cable would be needed? One of the first maps of the ocean depths was made by the American Mathew Maury in 1854 using only 200 measurements. When Maury drew his map of the North Atlantic he had no idea of how quickly it would become commercially important. The Atlantic Telegraph Company (ATC) had proposed a telegraphic link between Newfoundland and Ireland (over 3,000 km) and they asked Maury in the early 1850s whether such a link would be feasible. His reply was. 'The bottom of the sea between the places is a plateau which seems to have been placed there for the purpose of holding the wires of a submarine telegraph' (Clarke, 1992). With hindsight this was an overly confident assertion on the basis of 200 measurements.

The British and American governments provided ships for the cable laying. The Americans supplied one of their finest warships, the USS Niagara, and the British supplied HMS Agamemnon, a decrepit ship built just after the Napoleonic war and very close to retirement (see Figure 5.24).

Figure 5.24   HMS Agamemnon laying the first Atlantic telegraph cable in 1858. A whale is shown jumping over the cable.

After almost sinking in a storm the ships successfully laid the cable in 1858 across numerous mountain chains and valleys that Maury had never known about. The 'distance' from Europe to the new world had been reduced from weeks to minutes, creating an impact similar to the introduction of the internet 130 years later.

4.2 Motion in the ocean

tides

global ocean circulation

During a walk on any natural beach you will find debris (Box 5.5). This debris has been brought by two processes, a cyclical daily motion called the **tides**, and a longer term **global ocean circulation**.

Box 5.5 Drifting cargo

Most commercial goods are transported by container ship across the oceans. Even today, with detailed weather prediction, the largest ships can be damaged in storms (Figure 5.25). Sometimes containers are washed overboard and spill cargo.

(a)

(b)

Figure 5.25   The container ship APL China was damaged by a large storm in the Pacific Ocean and lost cargo valued at more than $100 million. (a) The ship, finally at port (b) A close-up of the damage.

In 1999 a container full of Nike training shoes was lost in the Pacific Ocean. Within a year these shoes were being washed up on the Pacific coast of North America. Soon newspapers were filled with adverts such as 'Nike Air Max size 8 left foot seeks right'. In the same way that computers can be used to predict the weather, they can also predict the circulation of the ocean. Figure 5.26 shows the predicted path the trainers would have taken.

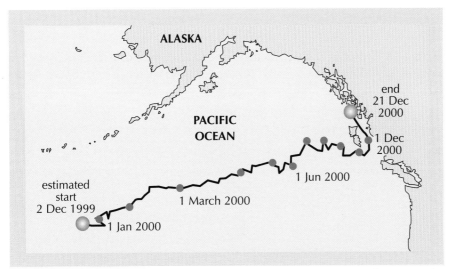

Figure 5.26 The predicted drift of Nike training shoes lost from a cargo ship across the Pacific Ocean.

Tidal motion

Most people are familiar with tidal motion, seen at any shoreline by the rising and falling of the water level. The forces that drive tides are the gravitational attraction between the Moon, the Sun and the Earth. These combine to raise and lower the sea surface by a few centimetres in somewhere like the Mediterranean Sea and up to several metres in somewhere like the River Thames in London. As these motions are cyclical they are not generally responsible for moving things about the oceans. If you throw some rubbish on a beach a high tide will wash it away. However it will most likely be brought back on the next high tide.

○   Could the tides be responsible for moving the Nike training shoes
         (Box 5.5) across the Pacific Ocean?

●   No, cyclical motion will not move the trainers across the Pacific.

Surface currents (wind-driven circulation)

Winds contain enough energy to move sailing boats (Figure 5.27). Some of this energy is transferred to the surface of the ocean through the action of friction between the wind and the sea surface.

Ocean waves grow in the winds and, in their own way, act as 'sails' to transfer energy from the wind to the ocean. This energy starts the oceans moving and typically, the wind-driven current is about three per cent of the wind speed.

Figure 5.27 The energy in the wind can be captured by the sail on a yacht to push it along.

Activity 5.11

If a wind of 10m s^{-1} was blowing over the surface of the ocean, what would be the typical wind-driven ocean current?

For the answer, go to the end of this chapter.

Three per cent may not seem much, but it is very significant. Anything that moves on the Earth is affected by the rotation of the Earth. The combination of the mean global wind pattern and the rotation of the Earth, sets up a pattern of surface currents shown in Figure 5.28.

○   Is it possible that the wind driven currents carried the Nike training shoes to the coast of North America from the central Pacific Ocean?

●   Yes. The region in which the training shoes were lost (see Figure 5.26), is around the southern edge of an anticlockwise circulation pattern in the north east Pacific Ocean. The trainers were most likely caught in this circulation and carried towards the North American coast. Closer to the coast the surface currents split and the trainers would have been carried in both the Alaska Current, and the California Current. These two currents would have deposited the trainers all along the Pacific Coast of North America.

The wind-driven surface currents are responsible for moving most marine pollution around the globe. Figure 5.28 can give an idea of where the impact of

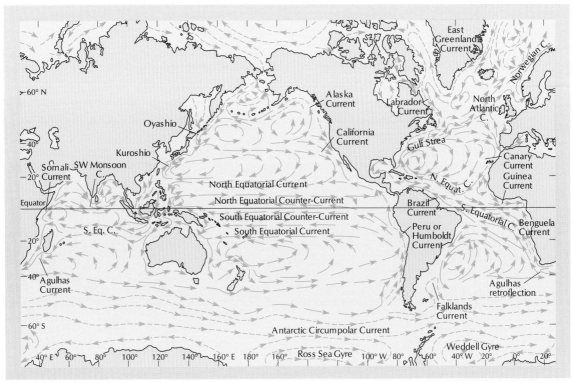

Figure 5.28 The surface currents of the oceans.

short term pollution could be. In Summer 2000 the Russian nuclear submarine Kursk sank in the Barents Sea (north of Norway) and radioactivity was soon detected along the paths of ocean currents north of Iceland.

Thermohaline circulation of the oceans

The final component of the ocean currents is due to the mixture of the surface currents caused by the winds and the climate of the Earth, called **thermohaline circulation.**

thermohaline circulation

Thermohaline circulation is water movement driven by differences in density of sea water (sea water density is a complicated function of temperature *and* salinity). The term thermohaline is derived from the temperature effects on the water (so 'thermo'), and effects due to different amounts of salt ('haline'). Sea water is heated and cooled at the Earth's surface by the Sun. Salinity is changed by evaporation, which increases the salinity (Section 2.2 above), and by fresh water from rain and rivers flowing into the oceans and decreasing their salinity.

To understand thermohaline circulation, imagine a swimming pool (Figure 5.29a). On the left hand side of the pool we cool the water, and on the right we warm it. When water is cooled its density increases and so the water sinks; where it is heated it becomes less dense and so rises (Figure 5.29b) Over time there will be an anti-clockwise circulation pattern as currents replace the water that sinks, and rises (Figure 5.29c).

We call this sort of process, where fluid movement is driven by external heat sources, **convection**. Convection can occur over small scales – such as in the smoke rising from a cigarette, through to our swimming pool example, and up to the driving of the circulation of the Earth's atmosphere (you will read about this in Chapter Six).

convection

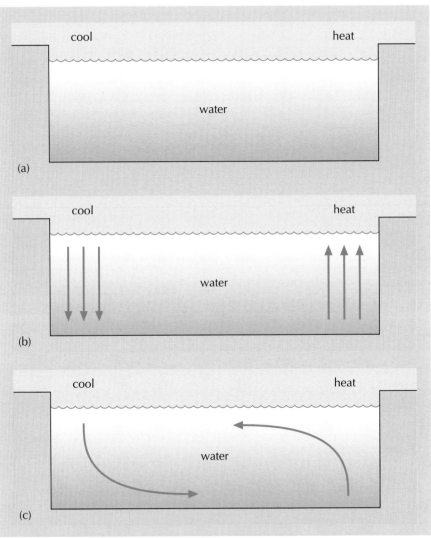

Figure 5.29 A hypothetical swimming pool showing the effects of temperature on the circulation of water. (a) The left side is cooled and the right warmed. (b) Cooled water sinks and warmer water rises. (c) After time there will be an anti-clockwise circulation.

Salt also affects the density of the water, the more salt, the higher the density. Over millennia, the heating and cooling, and addition and removal of salt has created a global ocean circulation pattern, shown in Figure 5.30. Although this is only a schematic, the cool area of the 'swimming pool' is in the Greenland Sea. Here the waters cool and become more salty through sea ice formation. This makes the water more dense, and so it sinks to the sea floor. These deep, cold, salty waters flow slowly southwards and finally enter the Indian and Pacific Oceans (blue path in Figure 5.30). These two oceans act as the warm end of the pool, and water density is reduced. The water rises and begins a long journey back towards the North Atlantic Ocean (orange/red path). This has been called the ocean conveyor belt and it may take several hundred years for water to go all the way around. You can see, by looking at Figure 5.30, that particular regions form the 'engines' that drive the conveyor only because of the shape of the ocean basins. In the Pacific Ocean water cannot get far enough north to cool and become dense. In the Atlantic the water moves to the polar regions relatively quickly.

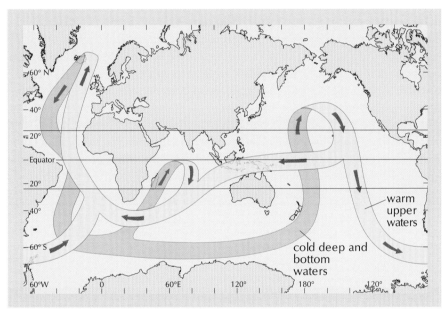

Figure 5.30 The ocean conveyor belt. Dense cold waters are coloured blue and they flow along the sea floor. Warm less dense red/orange waters flow near the surface.

As well as moving, the oceans provide a vast habitat for plants and animals. In the next section we shall investigate this.

Activity 5.12

If the climate of the Earth suddenly changed and the direction of the mean global winds (**wind fields**) changed, what would be the effect on the tides, the surface wind-driven currents, and the thermohaline circulation over a time scale of five years?

For the answer, go to the end of this chapter.

Wind fields are three-dimensional spatial patterns of winds.

4.3 Marine life

The ocean is a habitat for many plants and animals, from tiny plankton, through starfish to the largest animal to have ever existed on this planet: the blue whale. Life first evolved in the oceans and although we are familiar with some animals, there are some that we have not yet seen, and perhaps never will. Oceanic life has been relentlessly exploited by humans, sometimes sustainably but often not. As with life on land, oceanic life can be divided into different categories, which we shall discuss in this section. We shall also discuss two recent surprise discoveries that demonstrate that uncertainty (discussed in **Blowers and Smith, 2003**) is still an important concept: the Coelacanth and 'black smokers'.

The ocean as an environment

The differences between the properties of sea water and the properties of air make oceanic and land-based forms of life very different. In the first instance, the density of sea water is more than 1,000 times higher than air. This means that the animals and plants are, to a large extent, supported by the water, and it also takes relatively little energy for an animal actually to move (these two factors are partially responsible for allowing animals such as whales to become so big). Another property of sea water is that it strongly absorbs light. From Chapter One you remember that plants depend on light, and so they are limited to the near surface. Overall, oceanic life can be divided into two categories: things that live in the open ocean, the **pelagic** environment, and things that live on the sea floor, the **benthic** environment. Pelagic animals can live anywhere from the sea surface to the deepest depths. Benthic animals are restricted to the sea floor. Within the pelagic environment are two types of animals and plants: those that drift with ocean currents, called **plankton**, and those that can swim, called **nekton** (for example fish and aquatic mammals).

pelagic
benthic

plankton
nekton

○   Would plankton or nekton be more susceptible to changes in the surface circulation patterns?

●   Plankton are more susceptible as they cannot swim against the ocean currents.

The oceanic food web

phytoplankton

The oceanic food chain starts with small plankton called **phytoplankton**. Phytoplankton are plant cells that drift in ocean currents. They require light and minerals to grow, converting any energy from the Sun into chemical energy (see Chapters One and Two). Where plant nutrients are abundant, plankton rapidly grow and multiply, and despite their small size, the vast numbers mean they can discolour the ocean. Usually chlorophyll in plankton is not easily seen, however from space the picture is very different, and satellites can map the distribution (Figure 5.31). Plankton also includes small animals that are called zooplankton.

Figure 5.31 A satellite map of the distribution of plankton. Regions of plankton density are shown on a scale of 'hotness', the hotter the colour (more red), the higher the concentration of chlorophyll, and hence higher the plankton numbers.

There are areas in the global ocean where primary productivity is high (for example the North Atlantic). More surprising perhaps is how much of the ocean is the dark blue and mauve that indicates very low chlorophyll. In these areas, there are low levels of nutrients. Plankton convert solar energy to chemical energy, and where nutrients and light are abundant productivity can be high (red in Figure 5.31), with the opposite conditions productivity is low (blue/mauve in Figure 5.31). As this is the first stage in the food chain it is called primary production. The areas where the primary production is high are partially determined by the ocean currents that you read about above. They concentrate nutrients that promote growth in various parts of the ocean.

As you know from Chapter One, energy from primary production can be moved through a food web by herbivory and predation. Only a small proportion of the primary energy is transferred to **secondary production** at the next higher trophic level. In turn, when secondary production is consumed by larger animals energy is again moved to another trophic level. At each step only a proportion of the energy transferred is retained.

secondary production

Figure 5.32 shows part of a food chain with four trophic levels. From trophic level 1 (phytoplankton) to tropic level 2 (small plankton) only 20 per cent is converted to secondary production, 80 per cent is lost. Moving from trophic level 2 (small plankton) to 3 (large plankton), there is a further reduction and only a tenth of production is converted. In the final step, from large plankton to fish, again only one tenth of the remaining energy is converted. From our standpoint the food web is inefficient and when humans fish, we recover only a tiny fraction of the original production. This is an important factor when considering our exploitation of the oceans.

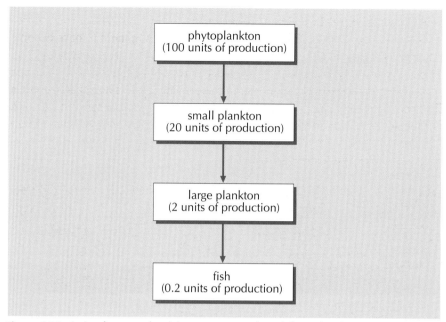

Figure 5.32 A simple example of the energy transferred through a food chain with four trophic levels.

Surprises in the ocean

hydrothermal vents

Recent oceanographic breakthroughs (for example, the discovery of deep-sea **hydrothermal vents** and a living 'fossil' in the ocean) can demonstrate to us the key concept of uncertainty.

Some sea floor sediments have a relatively high metallic content that is linked to the volcanic processes at mid-ocean ridges. It was not until the late 1970s that manned submersibles showed that the source of these metals were 'hot springs' on the sea floor. These hot springs, called hydrothermal vents, are part of the tectonic process that creates new sea floor. Cold water is sucked into cracks in the Earth's crust and is heated up. This hot water dissolves minerals in the crustal rocks turning the water into a very potent chemical solution that rises out of the crust in hydrothermal vents (Figure 5.33). Chimneys called 'black smokers' are formed around the heavily coloured chemical solution as it rises and cools.

Animals living at the edge of the smokers, in the chemical–rich hot sea water, form an ecosystem using hydrogen sulphide as its energy source instead of the Sun. The only other region in the oceans with high quantities of hydrogen sulphide is in the depths of the Black Sea where life is almost completely absent. The discovery of this new ecosystem has given us new hope of discovering life in other unlikely environments such as Antarctica and even other planets.

Figure 5.33   A black smoker (hydrothermal vent) on the sea floor.

Another 'surprise' was the catching of an unusual fish off the coast of South Africa in 1938. The fish was identified as a coelacanth species thought to have become extinct 70 million years ago; and yet here was one still alive! It is striking that, despite evolutionary pressure, one species can remain unchanged for 70 million years. The coelacanth has become famous as a 'living fossil'. The impact of such a discovery was huge, and when a second specimen was discovered in 1952, cartoonists used it to satirize the impact humans had had on the Earth. Figure 5.34 shows that in the 1950s there were also threats to the environment, such as the Korean War and nuclear weapons.

'If this is the best you can do in 50,000,000 years, throw me back.'

Figure 5.34 The cartoonist Illingworth makes his interpretation of what the Coelacanth would think of the world we created.

4.4 Human uses of the oceans

The oceans have always provided humans with resources and other benefits. Among the most important are food, a source of energy and minerals, a means of transport and a method for us to get rid of our waste. The global ocean circulation means that some of these uses could have global implications, and international consequences. In the late twentieth-century the recognition of these consequences led, in 1982, to the United Nations developing the Law Of The Sea (UNCLOS) as a means to solve some of the issues. This treaty gives maritime nations exclusive rights, up to 12 miles from their coastline, to do as they will. A further limit at 200 miles from the coast is called the **Exclusive Economic Zone** (EEZ). Here the legal status is different and other nations can, for example, transit at will, however, a nation still retains full rights to exploit the marine environment and sea floor up to this limit.

Exclusive Economic Zone

Food from the oceans

Fish is an excellent source of protein that, up until the middle of the twentieth century, must have seemed limitless. It has formed an important component in the human diet in many regions and is the only major exploitation in which humans are still acting as hunters. Almost 17 per cent of the world's requirements for animal protein is provided by the oceans and, globally, we eat on average approximately 13 kg of fish per person (FPP) each year. In the industrialized world this rises to approximately 27 kg FPP each year, with Japan consuming 72 kg FPP. In developing regions the consumption rate is approximately 9 kg FPP. Ocean productivity is not uniform (Figure 5.31) and over 90 per cent of the global fish catch occurs within 200 miles of land. However, only about 20 countries account for almost 80 per cent of the global catch.

In the 1930s the annual global catch of fish was under 40 million tonnes per year. In the second half of the twentieth century, production rose steadily until 1970, when it fell back before continuing to slowly rise to approximately 85 million tonnes per year (Figure 5.35).

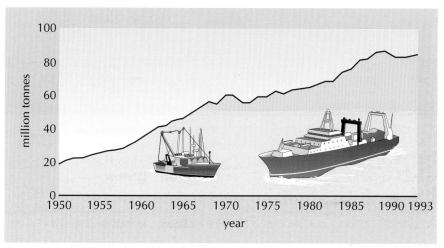

Figure 5.35 The trend of world total fish catch from 1950 to 1993 has been ever upwards.

Although huge, only approximately 5 per cent of all fish species make up this catch, and of this 5 per cent, 5 species account for just over half of the total global catch (pollock, pilchard, mackerel, anchoveta and tuna). The continued exploitation of these species could be acceptable if the exploitation is within the maximum sustainable yield (MSY). However, The Food and Agriculture Organization of the UN has estimated that the catch of 70 per cent of marine species has reached or exceeded the MSY and the total numbers of fish in the oceans will decrease. The industrialized world has discovered that people who live in cultures with fish as a key food source are healthier (for example the Inuit and Japanese) and so, predictably, these nations are increasing their fish intake. As 90 per cent of the fish catch is within 200 miles of coastlines, there is an issue over 'ownership' (which is covered by the UNCLOS) and European nations are now buying the rights to fish in the coastal waters of third world countries off the coast of Africa. Even if the world catch remained constant, the industrialized world is increasing its share. This increase can bring societies into conflict as highly mechanized industrial fishing, using factory ships (Figure 5.36a), kills many fish that are not eaten; local fisherman take only for their immediate needs (Figure 5.36b).

*The idea of a maximum sustainable yield was discussed further in **Drake and Freeland**, 2003.*

A solution for over-fishing could be **aquaculture**, in which fish are farmed in cages in the ocean, and harvested like other farm animals. Salmon are farmed in many Scottish sea lochs and this has increased the supply and reduced the price of this fish on the British market. However, there has been a great effect on the local ecosystem and the genetic variation of the total population has been reduced. Another solution could be harvesting at lower trophic levels. Figure 5.32 shows that by fishing at trophic level 4, we obtain only a fraction of the original production. If we decided to eat plankton then the amount of energy extracted from the ocean could increase greatly. However, if we remove energy at lower trophic levels this might have an impact on fish populations at trophic level 4 anyway.

aquaculture

Figure 5.36 (a) A factory ship. It has been estimated that almost 25 per cent of fish caught by these trawlers are discarded as waste. (b) Dominican fisherfolk. The artisan fisherman will only take enough for personal need, and local markets. This is almost certainly sustainable and such fishing has been going on for thousands of years.

Box 5.6 Fisheries quotas

Nation states control fishing with **quotas**. These quotas set strict limits on the amount of fish that can be caught in a year, and are set by governments or other international bodies. In simple terms, scientists estimate how many fish of a particular species exist and then determine the MSY (maximum sustainable yield). The problem is governments cannot only follow scientific advice: communities rely on the fishing industry to survive economically, for fish processing, ship repair and so on. Fish quotas are set to balance all of these priorities with sometimes disastrous results for the fish when the MSY is neglected. A recent example is the collapse of the Canadian cod fishery in the late 1980s: over a period of five years, inshore fishermen and experts warned the Canadian government that cod were being fished above the MSY and the population was in decline. When scientists reached consensus on the MSY for cod, it was immediately rejected because the reduction in fishing would damage communities too much. Predictably, within two years of the warnings (1992) the cod fishery had completely collapsed. As of 2002 the cod population had not recovered enough to allow sustainable exploitation, for reasons that are not fully understood. The Canadian financial loss has been estimated at one billion Canadian dollars. A similar path is apparently being followed by European governments in the North Sea cod fishery. Scientists have suggested catch quotas, and the European Union has rejected them as too severe. The World Wildlife Fund has responded by putting cod on the endangered species list and predicting that the North Sea cod fishery will collapse unless very drastic action is taken.

Energy and minerals

Energy can be extracted from the ocean, and resources on the sea floor can be exploited. There are currently two operational methods of extracting energy from the ocean and a third in development. The method extracting the most energy is from tidal barrages, which make use of the potential energy differences between high and low tides (see Figure 5.37). At high tide the barrage is opened and water floods into an enclosed estuary. As the tide turns the barrage is closed and the water is trapped. Water is then allowed to escape through turbines generating electricity just like river dams. Obviously the estuary will be greatly affected.

The second method extracts energy from waves. A float on the surface of the ocean will move up and down, and this can be used to generate electricity. A third method in development is called Ocean Thermal Energy Conversion or OTEC. Energy is recovered using natural temperature gradients between warm surface waters and deep cold waters. Although theoretically efficient there are serious practical problems and only experimental OTEC systems are operating.

Figure 5.37 The La Rance tidal barrage, France.

The most well known source of offshore energy is from oil beneath the sea floor. Our ability to extract oil has moved from water depths of tens of metres to hundreds of metres using large floating oil rigs on the Patagonian coast of South America, the Gulf of Mexico and the North Sea (Figure 5.38).

Minerals also exist on and beneath the sea floor, and scientists have shown that there are large areas with deposits of cobalt, nickel, copper and manganese. The practical extraction of these resources lies in the future.

Shipping

With the movement of passengers to airlines, it is easy to assume that the importance of ships in the global economy has diminished. The numbers of ships has decreased in the last 100 years, but ships are getting ever larger. In 1900 a cargo ship of 10,000 tonnes displacement was considered large. In 2000 there are ships fifty times larger, with oil tankers of over 500,000 tonnes displacement. With such huge ships there are issues both of wrecks and of the resulting pollution affecting the coastline.

Pollution and waste disposal

Pollution is a catch-all term that can mean many things to many people. The United Nations has defined it as 'The introduction by man, directly or indirectly, of substances ... which results or is likely to result in such deleterious effects as harm to living resources and marine life, hazards to human health, hindrance to marine activities, including fishing and other legitimate uses of the sea, impairment of quality for use of sea water and reduction of amenities' (United Nations, 2002).

Humans have always used the oceans to dispose of waste. A good example of how attitudes change is that, in 1958, a ship, the William Ralston, was scuttled in the North Atlantic. Deliberately sinking a ship is perhaps not big news but what made the sinking of the William Ralston significant was that it contained 8,000 tonnes of mustard gas, yet there was no public outcry. This can be compared with the furore about the attempted disposal of a floating oil storage facility – the Brent Spar – by Shell Oil in 1995. The publicity around the disposal was so bad that eventually Shell Oil disposed of the Brent Spar in Norway instead of the deep ocean. During the disposal it was found that the Spar contained 'only' approximately 50 tonnes of toxins.

Throughout history we have disposed of things in the ocean with no thought for future impacts. When the William Ralston was dumped, the deep ocean was thought to be a place where ocean currents would be very small. We now know this is not the case, and some of the most vigorous ocean currents are found at the sea floor. That said, we can now identify with reasonable accuracy the location of places in the ocean where substances could be dumped with only limited local effects. The question we have to answer is 'would such disposal be acceptable'?

Figure 5.38 An oil rig in the North Sea used to extract oil from the ocean depths.

The worst fears of environmentalists seemed to be realized when a large oil tanker called Braer was wrecked off the Shetland Islands in 1993 (Figure 5.39). The incident spilled over 80,000 tonnes of oil and Shetland's wildlife and local economy was thought to be in serious danger. Luckily some particularly bad weather broke up the oil slick and dispersed it. Long term effects have been surprisingly small although there has been an impact on the resident seabirds.

Although the impact of the Braer has been small, if another incident occurred it is unlikely that the same special circumstances would rescue the environment. For instance the wreck of the Sea Empress in South Wales in 1996 spilt almost 72,000 tonnes of oil with considerable impact on the local environment and wildlife.

Figure 5.39 The wreck of the Braer on the coast of the Shetland Islands.

Activity 5.13

If you had to dispose of nuclear waste in the oceans and you wanted to delay the return of this waste to the ocean surface, identify, by looking at Figure 5.30, where you think the best place could be for release of this waste.

For the answer, go to the end of this chapter.

Summary

There are three main ways in which the oceans can move; tides (which occur on a daily timescale and do not significantly move things around the ocean); second, global winds that set up surface currents that can move things on monthly timescales; finally, thermohaline circulation, which moves water around the Earth over timescales of hundreds of years.

As the largest environment on the planet, the oceans are very significant. Primary production is not distributed evenly across the global oceans and as energy is moved through the food web, much is lost. The oceans have provided two significant biological surprises in the last few decades: black smokers and the coelacanth. The coelacanth demonstrates that even with climatic changes, continental drift, and evolutionary pressure, some things can remain unchanged for 70 million years.

The ocean has many uses, most significantly as provider of food. Fish stocks have been overexploited and there are conflicts about future use and the estimates of MSY. The timescales for the recovery of fish populations are not known. The oceans provide sources of energy and minerals, but most are currently unobtainable and their potential for the future is again unknown. Pollution can be naturally dispersed or can create severe impacts both local and across national boundaries over time. Disposal of waste that can move about in the ocean is an international issue, as is the exploitation of fish stocks that move freely. Both require international treaties.

5   Chapter summary: a global commons?

The Earth is undeniably a blue planet that might more appropriately be called 'Water'. Water plays a fundamental role in life for all living organisms. This water is not 'fixed' in any one place and it moves around the hydrological cycle with time scales that vary from days to thousands of years. Fresh water is just a small component of this cycle, but it obviously has a key role for sustaining human life. We also exploit fresh water for many uses such as irrigation and power generation, but the quality of the water tends to decrease with use. We can artificially 'clean' this recycled water once the natural systems have broken down, but this takes both equipment, and money.

There is a huge global inequality in the use of water: would a person in the UK use 136 litres of water of water per day if they had to fetch and carry it themselves?

Figure 5.40 Water plays a fundamental role in life for all living organisms.

This use is not a problem if the water is from a sustainable source, but current global use is draining ancient aquifers that are not being replenished.

The ocean provides the Earth with resources: it acts as an environment and as a source of food and energy. It is not static, but moves on a variety of timescales and its resources are mostly finite. Humans have affected all of these resources through exploitation and human-induced climate change. For exploitation the timescales are relatively short. With renewable resources such as fish, the fact that the Canadian cod have not yet recovered demonstrates that we are still learning about biological systems, but the timescales are again decades. This means that depletion and extinction, rather than climate change, may be the critical issue in the next few decades.

The shorter residence time components in the hydrological cycle in Table 5.2 are more vulnerable to climate change. We know that precipitation will change and the atmospheric reservoir may increase. The net result of this change will affect the salinity of the ocean over longer time scales. The global wind field will also most likely change with the changing climate. Combined with changes in salinity, this will affect the global ocean circulation, the distribution of nutrients in the water, and most likely the distribution of biological activity (Figure 5.31). The timescale for these latter changes is however much longer – perhaps centuries.

In the ocean the UNCLOS defines national boundaries, but how we do we deal with fish and pollution that obviously ignores these limits? The UNCLOS suggests 'States have the obligation to protect and preserve the marine environment' (United Nations, 2002), which is of course the largest reservoir in the hydrological cycle. How to actually achieve this with the variety of competing spatial scales and timescales remains to be determined.

References

Blowers, A.T. and Smith, S.G. (2003) 'What do we mean by environment and why is it so important?' in Hinchliffe, S.J., Blowers, A.T. and Freeland, J.R. (eds).

Clarke, A.C. (1992) *How the World was One*, London, Victor Gollanz Ltd.

Drake, M. and Freeland, J.R. (2003) 'Extinction in time and space' in Hinchliffe, S.J., Blowers, A.T. and Freeland, J.R. (eds).

Hinchliffe, S.J., Blowers, A.T. and Freeland, J.R. (eds) (2003) *Making Sense of Environmental Issues*, London, John Wiley & Sons/The Open University (Book 1 in this series).

UK Climate Impacts Programme (UKCIP) (1998) *Climate Change Scenarios for the United Kingdom Technical Report Number 1*, Oxford, UKCIP. http://www.cru.uea.ac.uk/link/ukcip/Sum_Rep.pdf

United Nations Convention on the Law of the Sea (2002) *Oceans: the Source of Life* (1982-2002), United Nations.

Answers to activities

Activity 5.1

(a) London, around 600 mm annually, Cardiff and Belfast, between 800 and 1200 mm, Edinburgh around 800 mm.

(b) The average annual precipitation varies from less than 600 mm in East Anglia to over 3,200 mm in parts of Scotland, Cumbria and Wales.

Activity 5.2

The monthly precipitation in (a) Cambridge varies from 31 to 58 mm, and in (b) New Delhi from 2 to 165 mm.

Activity 5.3

Evaporation varies seasonally: it is greater in summer than in winter. The main reason for this is temperature: it is hotter in summer.

Activity 5.4

In the temperate maritime climate of the UK there is only a slight seasonal variation in precipitation (Figure 5.8) but a strong seasonal variation in evaporation (which is much higher in summer, Figure 5.9), so there is much more water lost to evaporation during the summer months than during the winter months. A hydrograph of a UK river therefore has a higher discharge in winter than in summer.

Activity 5.5

Globally, there is much more freshwater stored in lakes (0.025 per cent) than rivers (0.00007 per cent), which initially indicates that lakes should be the more useful freshwater resource. However, the residence time for water in rivers (a few weeks) is much less than for lakes (about 10 years), so river water has greater renewability, i.e. it is replenished faster in the river after it is abstracted for use than it would be in the average lake.

Activity 5.6

No reservoirs have to be built to store groundwater; it can be pumped from the ground as needed, avoiding the environmental side-effects of reservoirs. However, energy is used in pumping water from the ground, groundwater takes longer to renew than rivers (i.e. it has a larger residence time) and so, if demand outstripped renewability, groundwater would be an unsustainable resource.

Activity 5.7

Both Cambridge and New Delhi are in areas with less than the mean global precipitation and may need to store water for efficient use. The precipitation in

Cambridge has only a low seasonal variation whereas the New Delhi variation is highly seasonal, possibly needing reservoirs large enough to be able to store most of the year's precipitation when it arrives in the summer months. In contrast, significant precipitation occurs in Cambridge throughout the year and, therefore, there is less need for long-term storage of water.

Activity 5.8

(a) For the Aral Sea region, irrigation not only results in salinization, but also requires the diversion of water from rivers, resulting in the destruction of a productive fishery. This is therefore an inappropriate use of water.

(b) The irrigated agriculture is non-sustainable as the soil has become salinized.

Activity 5.9

Figure 5.21 shows that the area of increased summer precipitation is in north-west Scotland. This is the area of the British Isles with the highest precipitation, so an increase here would be less useful than if the summer rainfall increased in, say, south-eastern England.

Activity 5.10

The gradients on land are higher. We have road signs in Britain warning of gradients of up to 12 per cent – three times greater than we have in the oceans.

Activity 5.11

If the wind blowing on the surface of the ocean is 10 m s^{-1}, and the wind-driven current is three per cent of this, the typical wind-driven ocean current speed would be

$$= 10 \text{m s}^{-1} \times 0.03$$

$$= 0.3 \text{m s}^{-1}$$

$$= 30 \text{cm s}^{-1}$$

So the typical wind driven ocean current would be 30cm s^{-1}.

Activity 5.12

If the winds changed suddenly, over five years we should expect the tides not to change. This is because the forces that generate the tides derive from gravity and not the winds. The surface currents would change greatly: a new pattern of surface currents would be set up that reflected the new balance between the global wind circulation and the rotation of the Earth. The thermohaline circulation would not change at all over five years. The new wind field would have to be in place for many hundreds of years to change the schematic circulation shown in Figure 5.30.

Activity 5.13

Figure 5.30 shows that the cold water in the thermohaline circulation leaves the surface in the North Atlantic ocean. If we dumped nuclear waste here, we would expect that it would have to go all the way to the Pacific before returning to the surface on timescales of hundreds of years. However, we do not know the impact of radioactivity on benthic animals of the deep sea, and we do not know the way radioactivity would spread in the ocean.

Chapter Six

Dynamic atmosphere: changing climate and air quality

Roger Blackmore and Rod Barratt

Contents

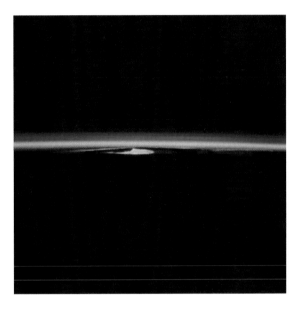

1 Introduction

A diverse assemblage of plant and animal life has thrived on Earth for millions of years, sustained by the Sun and supported by the soil, water and air. In previous chapters we have explored issues relating to energy, land and water; now we turn to the atmosphere.

The atmosphere is the envelope of air surrounding the Earth and is a global resource we often take for granted. The air we breathe circulates freely and the services it provides are freely available to all – to other species and ourselves. But, as with other Earth resources, the capacity of the atmosphere to absorb and recycle the waste products of modern society is not infinite.

In this chapter our focus is on changing climate: we outline some of the main global processes by which the dynamic atmosphere helps to create our climate, and then consider the effects that changes to the composition of our atmosphere will have on global climate. Like other chapters in this book we will seek to answer key questions:

- How and why are the atmosphere and climate changing over time and space?
- What are the consequences and significance of these changes?
- How can we represent and model these changes?

On completion of this chapter you will appreciate some of the spatial scales and timescales associated with atmospheric and global climate change. You will also appreciate some of the risks posed to human and natural systems by climate change, and understand how and why the impacts might increase with rising global temperatures.

2 The atmosphere

Like other planets, Earth receives energy from sunlight, but as far as we know, only ours has sufficient air and water to enable life as we know it to exist. In this section we describe some of the physical attributes of the atmosphere, starting with its structure and composition.

2.1 Troposphere and stratosphere

Compared to the size of the Earth, the envelope of air that forms the atmosphere is a thin shell (Figure 6.1). The part of the atmosphere we know best (called the **troposphere)** forms the lowest layer – varying in depth from approximately 15 km near the equator to only 8 km near the poles. This is the zone of weather and climate, a zone of continuous vertical and horizontal mixing of the air. It also contains the greatest bulk of the air because the atmosphere becomes less dense or 'thinner' as altitude increases. At 5,500 metres, approximately the height of the Tibetan Plateau, the **air pressure** is only half the value at sea level.

troposphere

air pressure

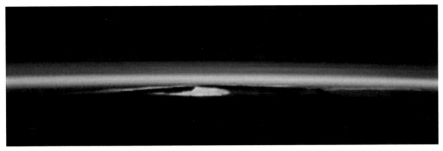

Figure 6.1 Photograph of the Earth's thin atmosphere taken from the Discovery orbiter vehicle. The atmospheric layers are evident.

Figure 6.2 Hot air balloon.

Air is a gas, and is therefore compressible. This means that, unlike a solid or a liquid, it can be squeezed into smaller volumes or allowed to expand into larger volumes. Atmospheric pressure occurs simply from the weight of the air above it and, because air is compressible, this means that air **density** increases towards the Earth's surface. We are not usually aware of air pressure, though if we were suddenly transported to somewhere with the altitude of Tibet we would find ourselves gasping for breath. Here we are mostly concerned with the effects that changes in air pressure have on density. The temperature, pressure and density of a gas are all linked. You may be familiar with the expression 'hot air rises'. This is because hot air is less dense than cool air – think of a hot air balloon (Figure 6.2). However, as an imaginary 'bubble' of hot air rises, the pressure of the air surrounding it falls, allowing the 'bubble' to expand. As a result, the temperature of the air in the 'bubble' will fall. Air temperature is normally highest near the Earth's surface, and declines with altitude. We call the rate of change of temperature with altitude the **lapse rate**. Its average value is 6 °C per kilometre in the troposphere (Figure 6.3).

Density is defined as the mass per unit volume. Think of a cup of feathers and a cup of sand. They both have the same volume (1 cup), but the sand will be much heavier and therefore is denser.

lapse rate

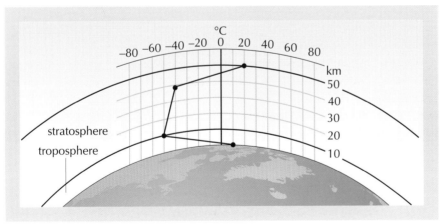

Figure 6.3 Vertical section through the atmosphere showing temperature gradients in the troposphere and stratosphere.

From the perspective of this chapter we will mostly be concerned with what is happening in the troposphere, our immediate environment, but we should not ignore the **stratosphere** (Figure 6.3). Here oxygen reacts with highly energetic radiation from the sun to form **ozone** at the same time producing heat so that in the stratosphere temperature *increases* with height. Stratospheric air circulation

stratosphere

ozone

is largely separate from that of the troposphere. This is fortunate because stratospheric ozone protects us from most of the Sun's damaging ultra-violet radiation while, at low altitudes, ozone is toxic to living things. It occurs naturally in very low concentrations in the troposphere, but is also a major component of much of today's air pollution.

2.2 Components of the atmosphere

The composition of the major components of the atmosphere has changed relatively little in recent geological time. Table 6.1 shows the typical composition of dry air. The proportions of major components (oxygen, nitrogen and argon) are nearly uniform up to approximately 80 km above the surface. Small amounts of other gases are also present. Note that the second column in Table 6.1 shows the *concentration* of each gas. Concentration refers to the mass or volume of something that is contained in a defined mass or volume of something else. In this case we use either percentage (parts per 100) or ppm (parts per million). According to Table 6.1, there are 78 units of volume of oxygen in every 100 units of volume of air, but just two units of volume of methane in 1,000,000 units of volume of air. The units of volume can be anything – for example cm^3 or m^3 – provided they are the same for both the air and the gas that is being measured.

Table 6.1   Approximate composition of dry tropospheric air

Gas	Typical concentration in air (volume/volume)
Essentially permanent	
nitrogen	78%
oxygen	21%
argon	0.9%
Variable	
methane	2 ppm
nitrous oxide	0.33 ppm
carbon dioxide	370 ppm
ozone	0.01 – 0.1 ppm
Very variable	
carbon monoxide	0.05 – 0.25 ppm
nitrogen dioxide	$(0.1 – 5) \times 10^{-3}$ ppm
ammonia	$(0.1 – 10) \times 10^{-3}$ ppm
sulphur dioxide	$(0.03 – 30) \times 10^{-3}$ ppm
hydrogen sulphide	$(<0.006 – 0.6) \times 10^{-3}$ ppm

Note: ppm = parts per million.

An **aerosol** is a collection of tiny liquid or solid particles dispersed in a gas, such as water droplets or dust in the atmosphere (inside or outside clouds).
particulate matter, see 'aerosol'.

In addition, **aerosols** (also known as **particulate matter**) are present, usually in low concentrations. There is one final component of the atmosphere we have not yet included; we have been considering the make up of *dry* air. The missing constituent is water vapour: molecules of water that behave like a gas and can vary in concentration in air from almost zero to 4,000 ppm.

You can now appreciate that we should not refer to 'pure air', for air is a mixture. The minor, or trace, components – gases and aerosols – vary considerably in space and over time, and these components may represent air pollution. Most have natural sources, but they can also come from human activities. When concentrations of these components increase to an extent that local or global air quality deteriorates, we can call this air pollution, defined by the World Health Organization (WHO) as:

> The presence in the outdoor atmosphere of one or more contaminants, such as dust, fumes, gas, mist, odor, smoke or vapor, in quantities, of characteristics, and of duration such as to be injurious to human, plant or animal life or to property, or which unreasonably interferes with the comfortable enjoyment of life and property.
>
> (WHO, 1980)

Although clean air should be freely available to all life forms, we humans have been gradually polluting it by releasing trace components, and in the process risking the health and well being of the entire planet. The scale of the problems associated with pollution has grown during the twentieth century, from short-term local pollution incidents through regional issues such as **acid rain** (see Chapter Five), to global-scale problems with long-term implications, notably **ozone depletion** in the stratosphere and global climate change.

acid rain

ozone depletion

Summary

The troposphere and stratosphere are two distinct layers of the atmosphere. The troposphere covers our immediate environment and is the zone of weather and climate. At increasing altitude within the troposphere, air pressure, air temperature and air density all drop. Air is made up of a mixture of gases plus aerosols and water vapour; changes to the components of air may lead to air pollution.

3   Solar energy, the atmosphere and climate

On a global scale, changes to the composition of the atmosphere are altering the climate system. To appreciate how this occurs we need to look at those patterns of natural energy flows in the atmosphere that create our temperate Earth and our weather and climate systems. We will consider each aspect in turn within this section.

3.1 Solar energy and the Earth

We are now familiar with sunlit views of the Earth from space (see, for example, Figure 5.1 in Chapter Five). Some of the features we see, notably clouds, snow and ice, appear bright because they reflect sunlight well. In contrast, forests, lakes and oceans appear dark because they absorb much of the sunlight and reflect little. The proportion of sunlight, or solar energy, that is reflected from a given surface is called the **albedo**. Those features with highly reflecting surfaces, such as clouds or fresh snow, have high albedos with values approaching 90 per cent, while those with poorly reflecting surfaces, coniferous forests and oceans for example, have low albedos with values typically between 5 and 15 per cent.

albedo

The planetary albedo is the combined figure for the Earth as a whole, and its time-averaged value has been measured as 31 per cent. This means that about one third of the energy arriving from the Sun is immediately reflected away to space, mostly by clouds, and is effectively lost to the Earth. The other 69 per cent is intercepted by the Earth and mostly absorbed at its surface. This remaining two-thirds is the amount of solar energy available to our planet and is an impressive quantity: the energy received *each day* from the sun is over 35 times greater than human world energy use for a whole *year* (Table 6.2).

Table 6.2   Comparison of solar energy and human use

(a) Solar energy intercepted each day	1.5×10^{22} Joules
(b) World energy use for the year 1997	4.2×10^{20} Joules
Ratio of (a) / (b)	1.5×10^{22} J / 4.2×10^{20} J = 36

When solar energy is finally absorbed (converted to other forms) at the Earth's surface, whether in the ground, in water, or by vegetation, it can be useful to know how much arrives at a particular surface over a given time period. This is the energy available per second, or power, across a unit surface area, and is measured in units of watts per square metre (W/m^2). When averaged over the whole Earth it has a value at the top of the atmosphere, that is before any reflection takes place, of 342 W/m^2.

○   What is the value of the rate of energy flow, measured in W/m^2, absorbed by the Earth's surface and atmosphere?

●   The value is 69 per cent of 342 W/m^2, or 0.69 x 342 W/m^2 = 236 W/m^2 approximately, roughly equivalent to two bright light bulbs shining non-stop for every square metre of surface.

3.2 The greenhouse effect and the temperate Earth

Over time, assuming the supply of solar energy it receives is steady, a planet such as the Earth will have a stable temperature so long as there is an approximate balance between the energy received from the Sun and that returned to space. Energy is emitted from the Sun mostly as **solar radiation**, and mostly in the

solar radiation

familiar form of the sunlight we can see and whose warmth we can feel. The Earth also emits energy, as radiation of the same electromagnetic energy group (Chapter Two), but with different qualities – we can not see it. The energy of this 'earth radiation' is transported by the same electromagnetic waves as is sunlight, but of longer wave length. We can make a distinction, then, between **short-wave radiation** from the sun and **long-wave radiation** from the Earth. The balancing act between short-wave and long-wave radiation for a planet without an atmosphere is illustrated in Figure 6.4a. Over a certain time period, a nominal 100 units of solar energy in the form of short-wave radiation reach the planet, and, assuming it has the Earth's albedo, 31 units will be reflected away. The remaining 69 units are absorbed at the surface, warming it. Balance is achieved when the planet emits the same amount (69 units of energy) out to space in the form of long-wave radiation, over the same period. The mechanism for this balance depends on some basic properties of materials outlined in Box 6.1.

short-wave radiation
long-wave radiation

Box 6.1 The radiative properties of materials

All bodies exchange radiant energy (energy in the form of **radiation**): they reflect, absorb and emit radiation. Reflection depends mainly on the nature of the surface, whereas absorption and emission depend on the nature and *temperature* of the surface.

Radiation, in general terms, means something that spreads out (radiates) from a source.

- Very hot bodies (e.g. the Sun) emit a range, or spectrum, of short-wave radiation, including ultra-violet, visible light and near infrared.

- Cooler bodies (e.g. the Earth) emit a range, or spectrum, of infrared or long-wave radiation.

- The hotter a body becomes the more radiant energy it emits from its surface.

○   Suppose the temperature of the Earth suddenly increased by a few degrees, what would then happen?

●   According to the last two points in Box 6.1, the hotter surface would emit more energy in the form of long-wave radiation. If the other energy flows (Figure 6.4a) are unchanged, the planet's energy budget is now in deficit and the surface of the planet would start to cool. Balance is returned when the surface of the planet returns to its original temperature.

Thus a planet like the Earth should have an equilibrium temperature that depends only on its albedo and the amount of short-wave solar radiation striking its surface over a given unit of time (known as *incident solar radiation*). This equilibrium temperature is known as the **effective radiating temperature**, which would be the same as the mean surface temperature of a planet with the orbit of the Earth but without an atmosphere. The effective radiating temperature for such a planet, can be calculated to be –19 °C.

effective radiating temperature

greenhouse gases

This is much lower than we experience at the surface of our planet because the Earth has an atmosphere containing **greenhouse gases** such as water vapour and carbon dioxide and these prevent the long-wave radiation emitted from its surface from passing directly into space. Instead, radiation from the surface is typically absorbed and re-emitted many times by the greenhouse gases in the atmosphere before some finally escapes to space, while some is returned to warm the Earth's surface (Figure 6.4b). The result is that long-wave radiation is finally emitted to space from an average level high in the atmosphere where the effective radiating temperature of –19 °C can be found. Because of the lapse rate (Section 2.1), the temperature near the surface is much warmer, a more temperate 15 °C averaged over the globe.

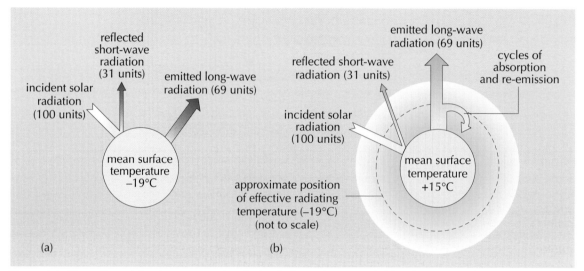

Figure 6.4 Illustration of the time-averaged radiation balance for an Earth-like planet (a) without an atmosphere and (b) with an atmosphere.

greenhouse effect

This atmospheric warming is called the **greenhouse effect** and the observed average surface temperature of the Earth is due to naturally occurring concentrations of greenhouse gases, such as water vapour and carbon dioxide.

Activity 6.1

Assuming a mean surface temperature of 15 °C and an average value of 6 °C per kilometre for the lapse rate, at what altitude would you expect to find a temperature of –19 °C?

For the answer, go to the end of this chapter.

While weather generally relates to day-to-day changes, climate refers to longer-term changes of weather patterns which are expressed in seasonality around the world.

3.3 Geographic distribution of climate types

So far we have looked at temperatures averaged over the whole Earth, but temperatures vary considerably with latitude, from the frozen polar regions to the hot Equatorial zone, because the Sun's energy does not fall evenly on the surface of the Earth. Imagine two identical beams of mid-day sunlight reaching the Earth's surface at the equator and at high latitudes (Figure 6.5). Near the Equator the mid-day Sun is high in the sky and the beam hits the Earth nearly full on, whereas at high latitudes the Sun is low in the sky because of the Earth's curvature, and the beam strikes only a glancing blow. Near the equator the beam of sunlight is spread over a smaller surface area, (Figure 6.5b), than at high latitudes (Figure 6.5a). So when averaged over a year, a given area near the equator receives more solar energy than at high latitudes, which is why tropical regions are warmer than polar regions.

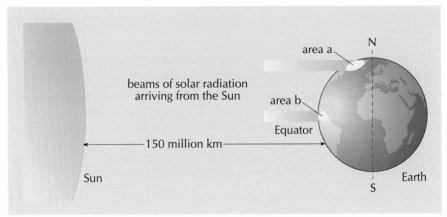

Figure 6.5 Two beams of sunlight reaching the Earth during the Spring equinox. The energy from the beam arriving at high latitudes is spread over area a, which is larger than area b, near the Equator.

Figure 6.5 shows the Earth during one of the two equinoxes when its axis is perpendicular to the Sun. At other times its axis tilts so that the angle at which sunlight strikes a given place on the Earth's surface also varies with the seasons. For half of the year the Northern Hemisphere is tilted towards the Sun, giving the northern summer and southern winter, and vise versa for the other half of the year when the Southern Hemisphere is tilted towards the Sun. This variation, both latitudinal and seasonal, in the amount of solar energy available at different latitudes on the Earth is the basis of our global climate system because it is the driving force for the large-scale patterns of air movement seen across the globe.

The process is illustrated schematically in Figure 6.6. The uneven distribution of heating at the Earth's surface creates a large-scale vertical air circulation (1), with air tending to rise near the equator and sink near the 30° latitude. This has the effect of transporting excess heat polewards, and leads to surface zones of equatorial low pressure and sub-tropical high pressure. This convection cycle is known as the Hadley Cell (see Chapter Five, Section 4.2, for a discussion of convection). When the effect of a rotating globe is added, airflows are rotated from a S–N direction to a more W–E flow, resulting in a surface pattern of latitudinal zones of high and low pressure, flanked by zonal air movements of westerlies and easterlies (2). (This idealized pattern represents average conditions, not what may be present on a given day.)

In response to the annual cycle of heating and cooling, more marked over the continents than the oceans, the pressure zones migrate north and south with the seasons, bringing seasonal rainfall to some parts. It is this seasonal march of temperature, rainfall and moisture availability (4) that characterizes the resulting climate zones (3). Starting at the equator and tracking polewards, we move from rainfall and water availability all season in the inner tropical zone to the summer (monsoonal) rains of the outer tropical zone, followed in turn by the dry zone of the sub-tropical high-pressure region. Continuing polewards we encounter a zone in which rainfall occurs mainly in winter, and then the all season rainfall of the westerly zone. Beyond this the climate is cold, and even though rain and snowfall may be limited, little evaporates so moisture is usually sufficient for cold-adapted organisms.

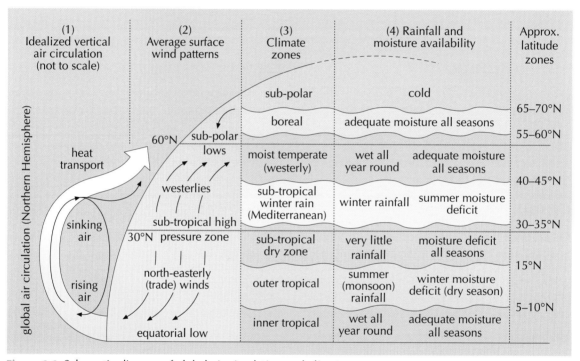

Figure 6.6 Schematic diagram of global air circulation and climate zones

Temperature and moisture form the basis for classifying biomes (see Figure 1.15, Chapter One), which map almost directly on to these climate zones.

Activity 6.2

Use the information shown in Figure 1.15 of Chapter One to complete the second column of Table 6.3 by inserting the appropriate climax vegetation types.

For the answer, go to the end of this chapter.

Table 6.3   A comparison of climate zones and climax vegetation types

Climate zone	Climax vegetation classification
inner tropical zone	
outer tropical zone	
sub-tropical dry zone	
sub-tropical winter rain	
moist temperate zone	
boreal zone	
sub-polar zone	

This account is much abridged. In reality the shape and distribution of large land masses complicates the picture. Large land masses are usually hotter in summer and colder in winter than the surrounding oceans, creating additional, seasonal air movements between oceans and continents. These monsoonal flows tend to bring more summer moisture to the eastern sides of continents than we have indicated. Nevertheless, you should still be able to track the journey we have described from the Equator polewards on the western side of the idealized continent described in Chapter One, Figure 1.15, and recognize the changes in climax vegetation described as we move through the different climate zones.

Summary

The atmosphere plays a vital role in modifying the incoming flow of energy from the Sun, balancing it against the outgoing radiation away from the Earth, and distributing it over the Earth. Naturally occurring greenhouse gases create the conditions for a temperate Earth. The uneven and changing distribution of solar energy at the Earth's surface shapes our weather and climate systems by driving the large-scale circulation patterns of the atmosphere which, in turn, determine the global distribution of climate zones.

4   Changing climate

Here are some quotes from an influential report, produced in 2001, by an international panel of leading climate scientists and other experts (IPCC , 2001a):

The Intergovernmental Panel on Climate Change (IPCC) was set up in 1988 by the World Meteorological Organization (WMO) and the United Nations Environmental Programme (UNEP).

- The Earth's climate system has demonstrably changed on both global and regional scales since the pre-industrial era, with some of these changes attributable to human activities.

- Globally it is very likely that the 1990s was the warmest decade, and 1998 the warmest year, in the instrumental record (1861–2000).

- There is new and stronger evidence that most of the warming observed over the last 50 years is attributable to human activities.

The same report also presents some new projections for future climate change and warns of likely impacts:

- ...an increase in globally averaged surface temperature of 1.4 °C to 5.8 °C over the period 1990 to 2100. This is about two to ten times larger than the central value of observed warming over the twentieth century and the projected rate of warming is very likely to be without precedent during at least the last 10,000 years.

- Ecological productivity and biodiversity will be altered by climate change and sea-level rise, with an increased risk of extinction of some vulnerable species.

These figures and comments have been taken up with relish by the media (Figure 6.7). Several of the newspaper headlines shown here give the impression that climatic disasters are about to occur.

Figure 6.7 Global warming and its impacts: the view from the media in 2001.

The instrumental record is the numerical data obtained from a variety of different instruments.

The aim of this section is to help you reach a better understanding of the extent, nature and significance of such changes to our climate, both current and in the future. To put these changes in context we start with a look at what has happened to climate and our atmosphere over a range of past timescales. Next we look at evidence that global climate is changing now and examine the reasons for these changes. Finally, we turn to projections of possible future climate change.

4.1 What do we know about recent climate change?

What is the evidence for recent climate change, and how sure are scientists that it is occurring?

Rising temperature

The key global indicator of recent climate change is rising temperature, specifically the increase in air temperature measured close to the surface by standard meteorological instruments. When averaged over the globe – over both land and sea – we call this the global mean surface temperature. Increases in this value are what are usually referred to as global warming. We will discuss briefly the trends of global temperatures over the last 140, 1,000 and 400,000 years.

Temperature changes during the last 140 years   Figure 6.8a shows the variations in global mean surface temperature from 1861 to 2000, which is the period from which global temperatures can be reconstructed using the **instrumental record**. Based on these data the estimate of the rise in global mean surface temperature over the last century is 0.6 °C ± 0.2 °C.

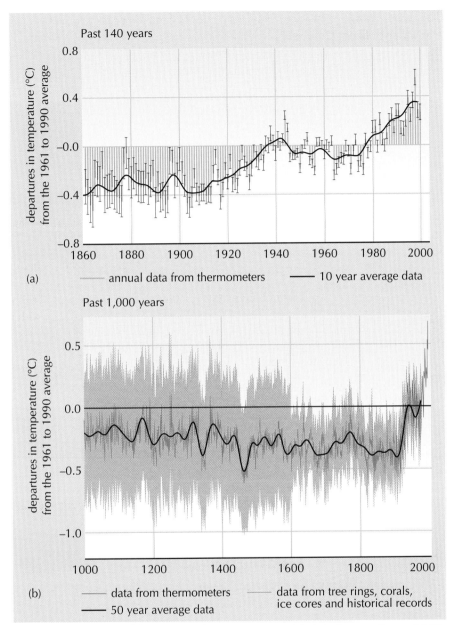

Past 140 years

(a)   —— annual data from thermometers   —— 10 year average data

Past 1,000 years

(b)   —— data from thermometers   —— data from tree rings, corals,
      —— 50 year average data   ice cores and historical records

Figure 6.8 Variations of the mean surface temperature (a) globally over the past 140 years, (b) in the Northern Hemisphere over the past 1000 years. In (a), error bars are attached to each individual year's data and don't always overlap the black line. *Source:* adapted from IPCC, 2001b, Figure 2.1.

How sure are climate scientists of these data? Uncertainty can arise for a variety of reasons: gaps in the data (for example, fewer observations over oceans where temperatures may have changed less than on land), the heating effects of urbanization, and the random errors associated with the instruments. Where the uncertainty can be estimated and quantified it is usually presented as a confidence range or interval, within which 95 per cent of the entire set of data points can be found. There is therefore a 95 per cent probability that any data

See **Bingham and Blackmore, 2003** for a discussion of probability calculations.

point will lie within this range. Thus, in Figure 6.8a the error bars, which are the lines extending above and below each data point, show the 95 per cent confidence range for each year's reconstructed temperature.

○   The 95 per cent confidence range also applies to the statement 'a mean temperature rise of $0.6\,°C \pm 0.2\,°C$'. Describe in your own words the information conveyed by this phrase.

●   The expected rise in mean temperature is $0.6\,°C$ and there is a 95 per cent probability that it lies between $0.4\,°C$ and $0.8\,°C$.

Temperature changes over the last 1,000 years   To estimate temperatures for the period before instrumental measurements were available, data are reconstructed from a variety of 'proxy' measures of temperature: historical records, tree rings, corals, sediments and ice cores (see Box 6.2). One such reconstruction is shown in Figure 6.8b. Because of the scarcity of proxy data from the Southern Hemisphere the mean surface temperatures are given for the Northern Hemisphere, not the whole globe. The grey region represents the 95 per cent confidence range in the annual data. You will see that these uncertainties are much larger than for the period during which we can rely upon the instrumental record, and uncertainty increases further back in time.

Box 6.2 Ways to reconstruct past climate

Humans have kept accounts of the dates of seasonal developments, such as cherry blossom in Japan and grape harvests in Europe, for nearly a thousand years. Records of notable climate events have been kept for even longer (e.g. droughts in Egypt). In addition a variety of techniques, ranging from counting annual growth rings (Figure 6.9) or microfossils to analysis of radio-isotopes, yields rich, if sometimes ambiguous, information from many biophysical sources.

For example, many kinds of trees experience an annual growth cycle, having a dormant or low growth period during winter or the dry season, followed by a growing season whose vigour is affected by temperature and moisture availability. When viewed in a cross section of the tree trunk, these cycles appear as concentric annual growth rings of narrow bands of dense, dark, slow growth wood, alternating with broader bands of less dense fast growth wood. The width and density of each annual growth band varies with local temperature and moisture conditions, creating a unique record that can then be matched with overlapping records to create longer time series. Thus, the annual growth rings of trees can provide information about temperature and rainfall conditions for each year from 1,000 years ago (and in a few cases to 11,000 years ago) to the present. Geographical coverage is good apart from tropical and subtropical regions where the absence of an annual cold period leads, at best, to poorly defined tree rings.

early, healthy growth   later, stunted growth   bark

one year's growth

Figure 6.9 Annual growth rings in a Black Forest fir tree, showing healthy growth, followed by twenty years of poor growth linked to acid rain pollution.

Similarly, in the oceans, cyclical seasonal responses lead to annual banding in corals, which can provide information about sea surface temperatures, sea level and other ocean conditions. Annual records typically go back 400 years, although in some cases fossil corals can give snapshots as far back as 100,000 years ago.

Layered sediments in lakes or the oceans are another rich source. The types of pollen trapped in lake sediments can reveal shifting patterns of vegetation and thus information about temperature and moisture. Records can go back millions of years, with each data point representing a period spanning around 50 years. In marine sediments, isotopes in microfossils can reveal data on temperature, salinity, atmospheric carbon dioxide and ocean circulation, while less regular deposits of sand can point to past dust storms and even to the break up of ice sheets with the release of detritus from melting icebergs. Marine sediments provide information from time periods ranging from 20 thousand to 180 million years ago, although older records cannot be as precisely dated as newer records.

The composition of ice and trapped air bubbles from continental ice cores in Greenland and Antarctica can also provide a wealth of information on historical sea surface temperature, volumes of continental ice, and atmospheric composition including greenhouse gases and aerosols. Annual layers allow accurate dating to about 40,000 years ago, but records go back 400,000 years with a resolution from decadal (data that represents time periods ranging from 1–10 years) to annual (Stokstad, 2001).

See Blowers and Smith, 2003 for more on the historical climate of the Northern Hemisphere.

Being able to reconstruct a broad-based, long-term data set is useful for climate scientists because it gives a picture of the natural variability of the climate over a long period. To mention one example, the existence in the Northern Hemisphere of a 'Medieval Warm Period' followed by a 'Little Ice Age' period has been widely commented upon. The data from Figure 6.8b suggest a slight cooling after the fourteenth century that lasted well into the nineteenth century. Climate changes during this period in the landmasses bordering the North Atlantic (N.E. America, Greenland, Iceland and north-western Europe) were often pronounced if not always synchronous, and are thought to have been associated with changes to the atmosphere and ocean centred on the northern North Atlantic. When conditions are averaged across the whole hemisphere, however, the changes no longer appear exceptional. In other words, this was apparently mainly a regional rather than a global phenomenon.

Natural fluctuations such as this are thought to occur all the time on regional and local scales and constitute the 'noise' or uncertainty against which climate scientists are trying to detect the global temperature changes. During the instrumental period (Figure 6.8a), for example, temperatures do not follow a consistent rise that might be expected from global warming, but instead fluctuate from decade to decade. Much of the scientific debate over the existence and causes of global climate change has centred on competing interpretations of such variations. The significance of the high temperatures of the last few decades is that they appear to rise above the range of natural variability observed over the thousand year period for the Northern Hemisphere. So it is likely, for this hemisphere at least, that the increase in surface temperature over the twentieth century has been greater than that for any other century of the last millennium.

Longer term changes   Ice cores taken from the Antarctic and Greenland ice caps provide a longer-term perspective by allowing climate record reconstructions that go back several hundreds of thousands of years. Beyond 10,000 years ago they paint a picture of an unstable climate oscillating between warm interglacial and cold glacial periods about every 100,000 years – with global temperatures varying by as much as 5° to 8 °C and associated sea level changes of 100 metres or more – interspersed with many rapid short-term fluctuations. Results from the Vostok ice core in the Antarctic (Figure 6.10) show the semi-regular oscillations of the glacial cycles. These are probably caused by cyclical variations in the Earth's orbit that slowly change the seasonal and latitudinal distribution of solar energy received by the Earth (although the mechanisms by which relatively small modifications to the energy budget are translated into glacial and inter-glacial cycles are by no means fully understood).

In contrast, global temperatures over the last 10,000 years have been much less variable, changing by little more than one or two degrees. This interglacial period appears to have provided the longest period of stable global climate for at least 400,000 years. It is almost certainly no coincidence that this is also when many human societies developed agriculture (Chapter Two) and when the beginnings

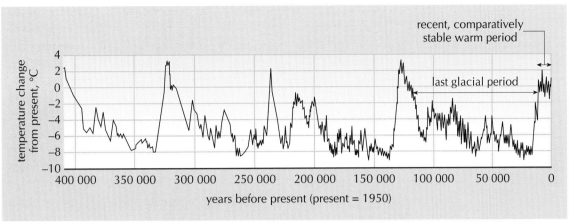

Figure 6.10 Temperature changes over the past 400,000 years reconstructed from the Vostock ice cores. *Source:* Petit et al., 1999, pp.429–36.

of civilisations (stable human settlements, irrigation and cities) occurred. It is ironic that today the waste products of modern industrial society may be ending this unprecedented stability.

Activity 6.3

From the perspective of the last 10,000 years, how significant are the sizes of the following temperature changes?

(a) The observed increase in the twentieth century of global mean surface temperature by approximately 0.6 °C (see Figure 6.10).

(b) The forecast for this century of an increase in global mean surface temperature of between 1.4 °C to 5.8 °C.

Other climate change indicators

So far, our review of past climates has focused on the variations of global temperature as the key indicator of climate change, but, as the box on proxy indicators shows, this is by no means the only component of the Earth's climate system to have altered. A great variety of other weather elements as well as physical and biological systems have been affected, and can also serve as indicators of recent climate change.

Physical and weather indicators of twentieth-century climate change The indicators selected here (Table 6.4) have been observed to change over large regions of the Earth during the twentieth century. According to the authors of the IPCC report, there is now a good level of confidence that what is being recorded is the result of long-term change rather than naturally occurring short-term fluctuations. We would thus expect the direction of movement of each indicator in Table 6.4 to be consistent with rising global temperatures.

Table 6.4   Twentieth-century changes in the Earth's climate system

Weather indicators	Observed changes
Hot days/heat index	Increased *(likely)*.
Cold/frost days	Decreased over most land areas during the 20th century *(very likely)*.
Minimum and maximum temperatures	Between 1950 and 2000 both increased *(very likely)*, but night-time minimum temperatures increased at twice the rate of daytime maxima *(likely)*.
Continental precipitation	Increased by 5–10% over the 20th century in the Northern Hemisphere *(very likely)*, although it has decreased in north and west Africa and parts of the Mediterranean.
Heavy precipitation events	Increased at mid- and high northern latitudes *(likely)*.
Frequency and severity of drought	Increased summer drying and associated incidence of drought in a few areas *(likely)*. In recent decades the frequency and intensity of droughts have increased in parts of Asia and Africa.

Physical indicators	Observed changes
Global mean sea level	Increased at an average annual rate of 1 to 2 mm during the 20th century.
Duration of ice cover of rivers and lakes	In mid- and high latitudes of the Northern Hemisphere decreased by 2 weeks during the 20th century *(very likely)*.
Arctic sea-ice extent and thickness	Thinned by 40% in recent decades in late summer *(likely)*, and decreased in extent.
Non-polar glaciers	Widespread retreat during the 20th century.
Snow cover	Decreased in area by 10% since satellite observations began in the 1960s *(likely)*.

Source: IPCC, 2001a
Note: The descriptions 'very likely' and 'likely' used in Table 6.4 and in the quotes from the IPCC on pages 241–2, for example, have been given a probabilistic interpretation by the authors. Specifically, 'very likely' means there is a greater than 90 per cent chance that a result is true, and 'likely' means there is a greater than 66 per cent chance. These levels of confidence are conservative estimates but are lower than the 95 per cent or 99 per cent usually used to indicate statistical significance in scientific literature. Generally in science a result is considered significant if there is less than five per cent ($P<0.05$), or in some cases less than one per cent ($P<0.01$), probability that the result occurred simply by chance.

The first three weather indicators clearly tell a story of rising temperature, with more frequent hot days and fewer cold events, although currently night-time temperatures are increasing more rapidly than during the day. However, the next three (changes to precipitation and drought) are less easy to interpret because they are very sensitive to any adjustment to the world's climate zones (see Section 3). Many climate model simulations now show climate zone shifts in response to warming, and precipitation changes of the kind described, but there is not yet sufficient agreement between them to be sure that the current changes are caused by a warmer world. We can say, however, that events of this nature would be expected from an increase in the rate of the hydrological cycle (see Chapter

Five). Warmer conditions give rise to increased evaporation and transpiration, and warmer atmospheres hold more moisture, thus hotter climates have a greater tendency both to seasonal droughts and to heavier precipitation.

○   Look at the changes in the physical indicators given in Table 6.4. How might these be explained by rising global temperatures? Are all the changes consistent?

●   The thinning and reduced extent of snow and ice over land and sea, and glacial retreat are all consistent with a warming of the climate. It is also likely that this warming has in turn contributed to the sea level rise, both through direct warming and expansion of seawater, and widespread melting of land ice. The changes are all consistent with rising temperatures.

Environmental Indicators   There is of course a wide range of other environmental indicators of climate change we could have included, for example biological and ecosystem responses. Here we should be wary of jumping to conclusions because we are dealing with complex systems that can respond to a great variety of other pressures and in unexpected ways. Nevertheless, there are a number of biological phenomena, such as animal migration and plant growth, flowering and fruiting, that cannot proceed until a minimum temperature has been reached over an adequate length of time. These **phenological** changes are easy to observe and monitor and can provide sensitive indications of climate change, and this is one field where amateur observers can make a valuable contribution, for example to national surveys. Quite a few studies have now reported changes from a range of regions and ecosystem types (Penuelas and Filella, 2001, pp.793–5). In the last 50 years, from Scandinavia to the Mediterranean and across North America, the growing season for plants has been increased by between one and four weeks. The leaves and flowers of most deciduous plants now appear earlier, but leaf colouring and fall is delayed (see Figure 6.11). Many animal life cycles also depend on temperature; for example in the UK it appears that aphids, which feed on young, tender parts of plants, now appear on average a week earlier than 25 years ago.

Phenology is the study of the seasonal timing of life cycle events, such as flowering and migration. Temperature is one factor that influences phenology.

Migrating animals, particularly butterflies and birds, also need to keep pace with the changes by arriving and breeding earlier in their summer habitat, so that food such as pollen and insects are available at the right time. Many are responding in just such a manner, whereas others are adapting by extending their ranges further north.

While there are many examples from weather and biophysical indicators that support the contention that global warming is occurring, caution must be exercised wherever records are of short duration, which in this context means less than a few decades. There are also some potential indicators of climate change that have yet to provide consistent evidence that can be distinguished from the natural fluctuations that occur from decade to decade. In spite of the

Figure 6.11 Although Spring has appeared earlier for trees in the Northern Hemisphere over the past 50 years, leaf colouring and fall has been delayed.

media coverage whenever a large storm hits an affluent country, no clear-cut global trends in the intensity and frequency of tropical and extra-tropical storms are yet apparent. Nor have significant trends been apparent in the extent of Antarctic sea ice during the comparatively short time (20 years) that accurate satellite observations have been available. A notable exception is the recent break up of some of its peninsular ice sheets in response to rapid regional warming (Figure 6.12), but it is too early yet to say whether this is simply a natural fluctuation or a precursor of more general warming.

One possible reason why the impacts of global warming have not yet been more damaging is that the buffering effect of the oceans has moderated the current pace of climate change. To understand this and to appreciate how our climate might develop in future years, in the following sections we review some aspects of the climate system and the causes of climate change, both human and natural.

4.2 The climate system

In Section 3.3 a simple climate model was presented: the uneven and ever changing distribution of solar energy at the Earth's surface drives the large-scale circulation patterns of the atmosphere that, in turn, determine the global distribution of climate zones. These large-scale air movements are also influenced by the surfaces they pass over, the oceans and their currents, the shape and nature of land surfaces, and the distribution of ice and snow. The atmosphere interacts with all the components of the Earth

Figure 6.12 The Larsen 'B' ice-shelf, on the Antarctic peninsula, just prior to its break-up in 2001.

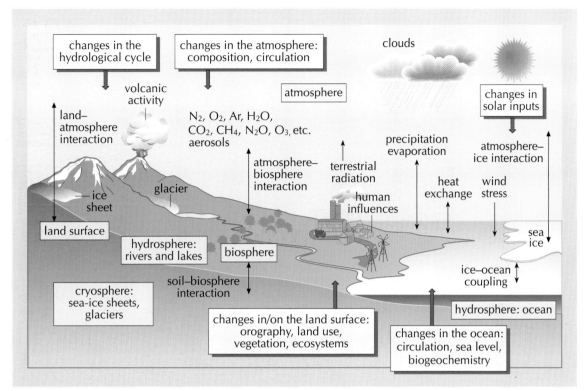

Figure 6.13 Schematic diagram of the global climate system illustrating the five realms of the Earth system (in blue boxes) and some of the interactions between them. Components and processes likely to contribute to change are in pale yellow boxes.
Source: adapted from IPCC, 2001b, Figure 1.1.

system and any comprehensive description of climate – which is needed to model climate change and its impacts – must take these into account (Figure 6.13).

In this more complex model the climate system responds to the five interacting realms of the Earth system: the atmosphere, the biosphere (life), the hydrosphere (oceans, rivers and lakes), the lithosphere (the Earth's crust and land surface), and the cryosphere (sea ice, ice sheets and glaciers). Together these processes determine the dynamic flows and circulation of materials and energy within the climate system.

Timescales

The processes indicated in Figure 6.13 take place over a range of timescales. One particular feature is of note: the dynamic atmosphere is the most rapidly changing component of the system; it changes from minute to minute, hour to hour. The others react and change more slowly. The oceans and ice sheets have their own much slower circulation patterns and flows, and it can take hundreds or thousands of years to build up an ice cap or warm the ocean throughout its depths (Chapter Five). The capacity of oceans to store large quantities of heat energy has a major influence on the climate system by slowing the rate of global warming.

However, this action merely postpones any warming and ensures that any changes will continue for a very long time. To take one example, the current rise in sea levels is mostly driven by thermal expansion from the warming of the upper levels of the oceans. The rise, at present, is comparatively slow but, even if air temperatures stabilize some time in the future, thermal expansion, and thus sea level rise, would continue for a millennium or more (Figure 6.14). This is because the warming will slowly continue to spread to all depths of the ocean, expanding an increasing volume of water. The melting of ice caps will further contribute to overall increases in sea levels.

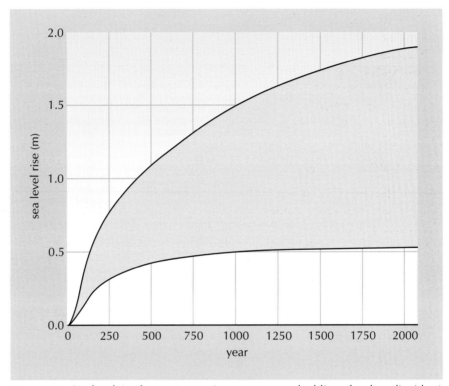

Figure 6.14 Sea level rise for 2000 years in response to a doubling of carbon dioxide. A range of results from different model runs is shown.
Source: adapted from IPCC, 2001b, Figure 11.15.

Ocean-atmosphere oscillations

The oceans thus act like the memory of the climate system and not only at the global level. Like the atmosphere, the oceans have their own global, regional and local circulation patterns, and all of these are capable of interacting with the dynamic components of the atmosphere. For example, unusually warm or cool regions of an ocean can often persist for many years and have a profound influence on the regional circulation of the atmosphere. Sometimes the dynamic interactions between atmospheric circulations and ocean surface temperatures can lead to more than one stable pattern, resulting in an oscillation between one preferred state and another.

Several of these ocean-atmosphere interactions have now been identified. The most well known is the **El Niño-Southern Oscillation** (ENSO), an interaction mostly in the Tropical Pacific, that alternates irregularly between its warm ocean state in the eastern Equatorial Pacific (*El Niño*) and its cold state (*La Niña*). In either of these two extreme states ENSO modulates the intensity and positions of many global weather systems leading to widespread droughts or floods in certain regions with pronounced environmental effects. On the time scale of two to seven years at which it operates, ENSO is the major source of variation in the climate system. Other oscillations with periods of a decade or more have been identified in both northern Pacific and Atlantic oceans. The **North Atlantic Oscillation** (NAO), for example, modulates the frequency of mild westerly winds across northern and western Europe. Its positive phase, which it has been in since the 1980s, is associated with mild temperatures, heavy rain and increased storminess across northern and western Europe, but drier weather in the Mediterranean. These interactions are more than scientific curiosities, they have profound influences on the regional climate, are a major source of natural variability in the climate system and may themselves be changing in response to global warming.

El Niño-Southern Oscillation

North Atlantic Oscillation

Feedbacks

Another feature of interactions within the climate system is that they often exhibit feedback, which can make the outcomes more difficult to determine. Feedback can occur in a system when the interactions between its component processes respond to a change of conditions. The interactions can either augment the original change (positive feedback) or reduce it (negative feedback). The general process is shown in the schematic diagram below (Figure 6.15).

Chapter Three provides a discussion of similar interactions affecting populations.

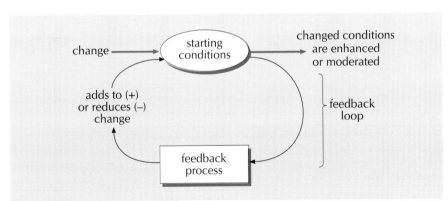

Figure 6.15 A schematic diagram illustrating the general effect of both positive and negative feedback.

Feedback is easier to understand using specific examples; we will use two that occur within the climate system. In recent decades the high latitudes of northern land masses have been warming at several times the global rate. One reason is that warmer temperatures are melting the winter snow cover earlier. Bare ground absorbs much more of the Sun's energy than snow, leading to an additional heating effect ... which then helps to melt more snow. This is an example of positive feedback (see Figure 6.16).

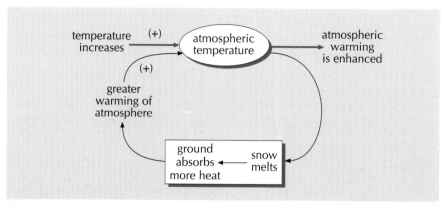

Figure 6.16 A schematic diagram using snowmelt to demonstrate positive feedback.

Negative feedback can be illustrated by a side effect of increasing the concentration of carbon dioxide in the atmosphere, which is to increase the rate of growth of many species of plants (known as carbon dioxide fertilization). The rapidly growing plants take up more carbon from the atmosphere and moderate the increase of carbon dioxide in the atmosphere (see Figure 6.17). In spite of the connotation of the word 'negative', the effect of negative feedback is to reduce the perturbations in a system and thus to stabilize it.

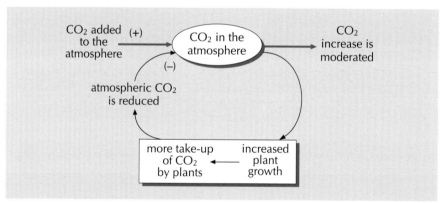

Figure 6.17 A schematic diagram using 'carbon dioxide fertilization' to demonstrate negative feedback.

4.3 The causes of climate change

In Section 3.2 we saw that the natural greenhouse effect currently maintains the global mean surface temperature at approximately $15\,°C$, a value that has changed little over the past 10,000 years. Since the industrial era, however, human activities have increased emissions of a variety of pollutants into the air, and this has led to the accumulation of carbon dioxide and other greenhouse gases in the atmosphere. The concern is that this will lead to global warming and climate change on a much greater scale than has already been experienced.

Carbon dioxide in the atmosphere

The evidence of changing atmospheric concentrations of carbon dioxide comes from direct measurements in the last 50 years and a variety of proxy sources that trace its fluctuations over longer periods. Figure 6.18 shows these changes on four different timescales.

From Figure 6.18b we can see that carbon dioxide concentration remained stable over the last millennium at approximately 280 parts per million (ppm), until about 1750 when it started to rise exponentially. Figure 6.18a shows the Mauna Loa data (taken from a Hawaiian mountain peak in the Pacific Ocean) in more detail, charting the continuing rise of the last 50 years, which reached a value of 368 ppm by 2000. These recent levels (which have been added in green to the two longer sequences of Figures 6.18c and 6.18d) are in stark contrast with the fairly constant composition of the last ten thousand years (Figure 6.18c), which marks

Exponential increase is described further in Chapter Three.

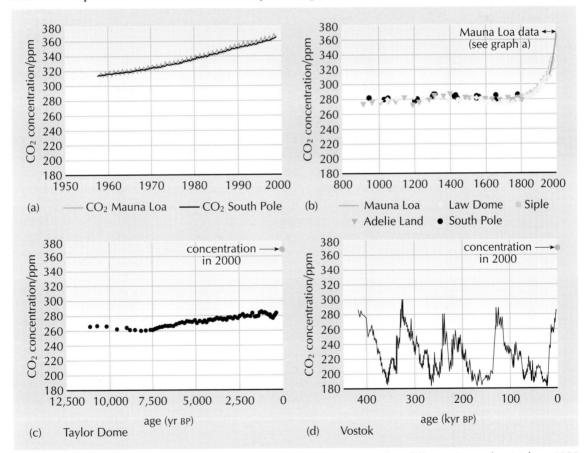

Figure 6.18 Variations in atmospheric carbon dioxide concentrations on four different timescales: (a) from 1958 to 2000 (b) from 800 to 2000 (c) from 12,500 years BP to present day (d) from 400,000 years BP to present day. The rising trend from Mauna Loa, detailed in Figure a, is also shown (in green) in Figure b.
Source: adapted from IPCC, 2001b, Figure 3.2.

the comparatively stable climate of the Holocene. Figure 6.18d shows the cyclical variations of carbon dioxide concentration, which move approximately in concert with the rise and fall of temperature during the glacial cycles that we encountered earlier (Figure 6.10).

○   Compared to the values seen in Figures 6.18c and 6.18d, is the recent rise in concentration of carbon dioxide to 368 ppm unusual?

●   It certainly is. None of the earlier periods, going back several hundred thousand years, experienced carbon dioxide concentrations any higher than 300 ppm, even though they include very significant changes of temperature of the glacial and interglacial periods.

Recent projections (IPCC, 2001c) have suggested that, in the absence of concerted action, carbon dioxide concentrations will increase to between 540 and 970 ppm by the year 2100. Although this range is very wide there can be no doubt that the changes we are creating in the composition of the atmosphere are unprecedented. The consequences arise mainly from changes to the **greenhouse effect**.

greenhouse effect

The enhanced greenhouse effect

You will recall from Section 3.2 (Figure 6.4a and b) that over time a balance is maintained near the top of the atmosphere between incoming solar radiation and outgoing long-wave radiation. Any factor that alters either the incoming or the outgoing radiation and changes this balance is called **radiative forcing,** and has the units W/m^2 (watts per square metre). A positive radiative forcing tends to warm the Earth's surface and the lower atmosphere, while a negative forcing cools them. If, for example, the Sun's output increased it would lead to a positive radiative forcing, whereas if it decreased a negative radiative forcing would occur.

radiative forcing

The net effect of recent human influences on the climate system (Figure 6.13) has been to increase radiative forcing, primarily by adding greenhouse gases to the atmosphere. This addition to the existing influence of natural concentrations of greenhouse gases is known as the **enhanced greenhouse effect**, and the main contributor is carbon dioxide.

enhanced greenhouse effect

Suppose now that carbon dioxide in the atmosphere is doubled but that the rest of the climate system remains the same. In this case, climate models agree that the outgoing radiation, currently $235\,W/m^2$, would be reduced by about $4\,W/m^2$, or a little under 2 per cent.

○   Would you expect the radiative forcing to be positive or negative?

●   The forcing would be positive. Incoming radiation stays the same, but outgoing radiation has been reduced, so the net balance is increased by $4\,W/m^2$. The effect is as if the Sun's output had been increased.

To restore the balance the same models estimate that the temperature of the surface/lower atmosphere would have to increase by approximately 1.2 °C ± 0.1 °C. In reality, the response of the climate system is much more complex because feedbacks may amplify the increase to between 1.5 °C and 4.5 °C (refer back to Section 4.2 and feedbacks). The most important contributor to the amplification of the temperature increase is water vapour, which also has greenhouse properties. This is because the capacity of the atmosphere to carry water vapour increases almost exponentially with temperature, giving rise to a strong positive feedback cycle between a warming near the surface, more water vapour and an increased greenhouse effect. A significant part of the scientific uncertainty in modelling climate change is due to our limited knowledge of these feedbacks and also of clouds and their interactions with radiation.

Activity 6.4

Is the water vapour feedback described above an example of positive or negative feedback? Complete the schematic diagram of the mechanism shown below, then compare your diagram with the answer at the end of this chapter.

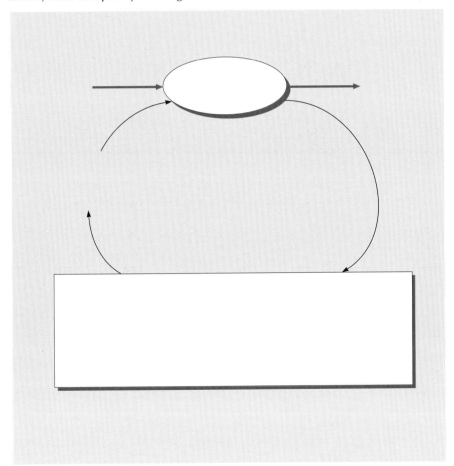

Figure 6.19 A schematic diagram showing water vapour providing positive feedback to atmospheric warming.

Climate change agents

We have already noted that in addition to carbon dioxide there are other trace components of the atmosphere (aerosols as well as gases) that can affect the radiation balance of the climate system and have the potential to cause climate change. These components can be classified into three categories:

- well-mixed (long-lived) greenhouse gases, notably methane, nitrous oxide, most halocarbons
- short-lived greenhouse gases, notably tropospheric ozone
- aerosols of many types, from a variety of sources.

In addition there are other agents of change not located in the atmosphere, the two most notable being the alteration of the albedo or surface reflectance of the Earth by changing land use, and variations in the output from the Sun. While many of these factors or **climate change agents** (Table 6.5) have natural origins, human activity has been responsible recently for adding to the atmospheric load of greenhouse gases and aerosols. Between 1750 and 1998 carbon dioxide has increased (Figure 6.18b) by 31 per cent, methane by 150 per cent and nitrous oxide by 16 per cent (IPCC, 2001a).

Recent increases in greenhouse gases have added to the 'natural' greenhouse effect to increase global warming (positive contribution) while increases in aerosols have tended to have the opposite effect (negative contribution).

Table 6.5 lists the agents of climate change that are thought to have contributed to recent global climate change and summarizes some of their properties.

Activity 6.5

Refer to Table 6.5. What would be the effect on the climate system's radiation balance of suddenly stopping all emissions due to human activity of (a) carbon dioxide, and (b) sulphate aerosols?

For the answer, go to the end of this chapter.

The cumulative effect of these agents for the period from the pre-industrial era, 1750, to the year 2000 is presented in Figure 6.20, with the contributions expressed in terms of radiative forcing in W/m^2.

In addition, Figure 6.20 estimates the level of scientific understanding that accompanies each estimate. Only for the well-mixed greenhouse gases and ozone is confidence high, and for several categories the error range, represented by vertical lines, is too large to be certain of the direction of the effect. This diagram strongly implicates greenhouse gases in the recent global warming, but the effect of other climate change agents is still very uncertain. A few scientists believe that the development of global warming in the last century has been caused by solar variations not greenhouse gases, although Figure 6.20 suggests that this is unlikely.

Halocarbon is a general term covering CFCs (chlorofluorocarbons) and similar compounds containing carbon, fluorine, chlorine and bromine.

climate change agents

Table 6.5   Climate change agents

Type of agent	Sources of change	Contribution to recent global warming	Persistence of effect
Well-mixed greenhouse gases: carbon dioxide, nitrous oxide, methane and CFC family	Natural + fossil fuel use, biomass burning and agriculture, and some more specialized uses for CFCs that are entirely human made	Positive contribution	From ten to hundreds of years or more
Tropospheric ozone	Photochemical reactions involving methane and other pollutants, e.g. from traffic exhausts	Positive contribution	Hours to days
Stratospheric ozone loss	Natural + Interactions with CFCs and other halocarbons	Mostly negative contribution	Months
Aerosols: sulphates, soots and mineral dust	A variety of natural and human sources depending on type: fossil fuel and biomass burning, mining and industry, land use change	Mixed – both positive and negative contribution depending on type; the net effect is likely to be negative	Days to weeks
Volcanic activity	Aerosols injected into the stratosphere	Negative contribution	Months to years – intermittent
Land use (albedo changes)	Natural and human, deforestation (and afforestation) is a major factor	Mixed – mostly negative	Months to decades
Solar influences	Natural	Probably positive contribution, but on longer time scales mixed – both positive and negative	From decades (the solar sunspot cycle) to up to a 100,000 years (Earth orbital changes)

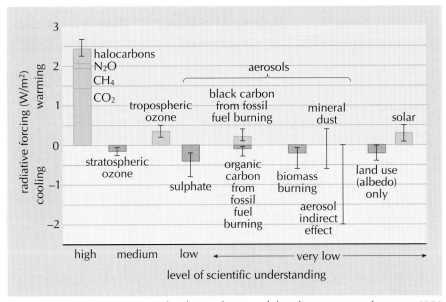

Figure 6.20 Climate agents and radiative forcing of the climate system between 1750 and 2000.
Source: adapted from IPCC, 2001b, Figure 6.6.

Activity 6.6

How does the information given in Figure 6.20 justify the point made in the last sentence – that it is unlikely that recent global warming can be explained by variations in solar radiation.

4.4 What can climate models predict?

Climate models simulate, mathematically, the working of the climate system and can be used to investigate the behaviour of past and current climates, and estimate the state of future climates. Since their development in the 1970s to model the behaviour of the atmosphere, general circulation models (GCMs) have become increasingly sophisticated in modelling climate processes and their interactions with climate change agents. They now model not just the atmosphere but the dynamics of other systems with which they interact. A major breakthrough was coupling the circulation of the atmosphere to that of the oceans, because oceans have a major influence on the fluctuations and pace of change of climate over any but the shortest timescales, as we saw in Section 5.2. Coupled GCMs now provide credible simulations of climate on spatial scales from the global to sub-continental, and on time scales from seasons to decades.

This has led to increased confidence in model projections for future climate change, and to the forecast quoted earlier that 'Globally averaged surface temperature is projected to increase by 1.4 °C to 5.8 °C over the period 1990 to 2100'. But here we face a contradiction: how can such a wide range of uncertainty accompany increased confidence? To answer this it helps to be aware of the stages involved in modelling future climates.

Stages in climate modelling

Stage 1 A set of scenarios is developed for the future state of the world's societies and economies, then used to estimate future population, economic activity, and use of technology, energy and other resources.

Stage 2 Predictions are then made from each scenario of future emissions of greenhouse gases and aerosols (and other impacts such as changes in land use). These are used to calculate future concentrations of these components in the atmosphere and hence the change in energy balance in the atmosphere, the radiative forcing.

Stage 3 The changed atmosphere is then used to predict overall changes to the climate system: from weather elements such as temperature and precipitation to biophysical components such as the state of the oceans, ice and vegetation cover.

Stage 4 The new state of the climate system can then be used to investigate the likely impacts for environmental systems, resources and human societies.

It is clear there are many assumptions and uncertainties in trying to predict future human behaviour and its possible influences on climate change agents. The associated uncertainties are at least as large as the scientific uncertainties arising from the climate models. Because predictions of future climate change depend critically on the socio-economic scenarios of the first stage, the term **projection** is preferred to prediction.

In this context **projection** refers to a prediction based on a specific set of social and economic assumptions.

The recent projection of a global temperature rise 1.4 °C to 5.8 °C over the period 1990 to 2100 (IPCC, 2001a) has rightly attracted media attention because the range is much wider and the upper value is significantly higher than in earlier projections. But changes in the climate model calculations are not the cause; they are due in the first case to the adoption of a wider range of scenarios in stage 1, and in the second to some scenarios reducing their projections of sulphate aerosol emissions.

○ Why might cleaning up the atmosphere by reducing sulphate emissions lead to higher global temperatures?

● Reduced emissions leads to reduced concentrations of sulphate aerosols in the atmosphere, which *removes* a cooling effect. The net effect of aerosols on global warming is negative: Table 6.5. So, perversely, cleaning up this source of pollution adds to global warming and its effect is immediate.

Climate model projections

Climate models are thought to represent well the global distribution of future changes in temperature, because similar results are achieved by a range of different models. Figure 6.21 shows two typical projections: the expected changes between the twentieth century and the end of this century (2071–2100), for two recent middle of the range scenarios, A2 and B2 (IPCC, 2001c), which project global mean surface temperature changes of 3 °C and 2 °C respectively.

○ Describe the main features of the distribution of temperature change.

● Nearly all land areas warm more rapidly than the oceans, Arctic Ocean excepted, particularly at high northern latitudes

Projections of changes in precipitation by different models also show some consistent themes, in particular increased precipitation at high northern latitudes. Elsewhere the results are less clear and sometimes contradictory largely because, when compared to temperature, precipitation is generally much more variable from one time period to the next, or from one location to another. When the combined effects of projections for increased temperature and changed precipitation are put together to estimate changes to water availability, confidence in predictions at a regional level is not high.

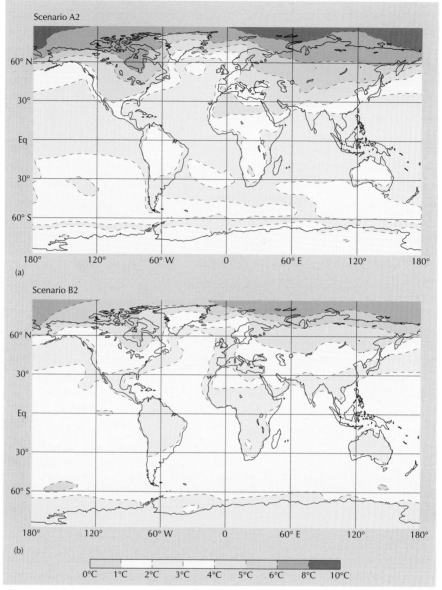

Figure 6.21 Two projections of annual mean change of temperature for the end of the 21st century (a) Scenario A2 (b) Scenario B2
Source: adapted from IPCC, 2001b, Figure 9.10.

○   Why is the balance of available water in a region significant?

●   It is critical for all life – it plays a major role in biome distribution and agriculture, and plants, in their turn, dictate to some extent the distribution of other species (including humans).

Summary

Over the past century, global mean surface temperature is estimated to have risen by 0.6 °C ± 0.2 °C. Long-term comparisons of temperature change are necessary before we can determine whether or not this increase is in line with natural temperature variation. Data from ice cores show that global temperature has changed little over the past 10,000 years, suggesting that the increase over the past 100 years is unusual. Other factors such as drought and increased rainfall in some areas, plus phenological changes in numerous species, point to a twentieth-century climate change. Climate modelling can help us to predict future change. Such models are becoming increasingly sophisticated although they are complicated by processes such as feedback and socio-economic changes.

5   The significance of climate change impacts

The ultimate objective of the UNFCCC (United Nation Framework Convention on Climate Change) is to 'prevent dangerous [human] interference with the climate system'. In the context of some fairly robust science but with models still showing large ranges of uncertainty in their projections, what exactly do we mean by dangerous? One tentative conclusion we have reached is that the recent global increase of 0.6 °C might present problems for some species in some areas but provide benefits to others. But what is the effect of an additional global change of 3 °C, which is about at the mid-range for climate projections at the end of the century? A statement such as 'a 3 °C rise in global mean surface temperature' tells us no more about the impacts of climate change than knowing that the average global temperature of 15 °C tells us about the climate in Moscow or Mombasa. We need to investigate the effect that global climate change will have on individual regions, and explore the impact on *local* ecosystems, resources and societies of the interplay of weather elements, such as precipitation and temperature, and physical elements such as sea level rise. The range of systems and sectors affected is broad: hydrology and water resources; agricultural and food scarcity; terrestrial and fresh water systems; coastal zones and marine ecosystems; human health; human settlements, energy and industry; and insurance and other financial services. Those that are most at risk of harm from climate change may vary from region to region and, accordingly, in the following discussion most of our examples will focus on specific regions and sectors.

5.1 Sensitivity and vulnerability to climate change

In trying to assess the impact of climate change on a wide variety of systems, we encounter several challenges. The responses are not always well understood. The impacts on a given system are often hard to distinguish from changes and pressures from other sources. In addition, and we are comparing different types of risk, for example high-probability low-hazard impacts with low-probability high-hazard impacts (**Bingham and Blackmore, 2003**).

sensitivity

vulnerability

The approach we take here has two strands. First, we distinguish between the sensitivity of physical, ecological and social variables to climate change, and their vulnerability (IPCC, 2001d). **Sensitivity** in this context is simply the degree to which some feature of an organism or system is affected by a given climate-related stimulus. Some plants and crops may be very sensitive to an increase in maximum temperatures, leading, for example, to a sharp reduction in crop yield above a certain threshold, while others could easily tolerate, or even welcome higher temperatures. **Vulnerability** is the degree to which that organism or system is unable to cope with the adverse effects of climate change, including variability and extremes. It can depend on the nature, size and speed of the changes, the sensitivity of the system and its capacity to adapt by moderating or even taking advantage of the changes. For example, in rich nations, agriculture as a whole is unlikely to be vulnerable since farmers can usually afford to pay for different varieties of crop, or change crops. Farmers in poorer nations may not have these options, and many ecological systems may literally have nowhere to go: both are more vulnerable.

Second, we identify some of the different types, or dimensions, of the hazards that can arise from climate change and are major causes for concern. In each of the following four sections, a different type of climate hazard is identified and discussed, using a specific example.

Hazards to unique and threatened systems

According to the IPCC authors (IPCC, 2000d) 'Natural systems can be especially vulnerable to climate change because of limited adaptive capacity, and some of these systems may undergo significant and irreversible damage. Natural systems at risk include glaciers, coral reefs and atolls, mangroves, boreal and tropical forests, polar and alpine ecosystems, prairie wetlands, and remnant native grasslands. While some species may increase in abundance or range, climate change will increase existing risks of extinction of some more vulnerable species and loss of biodiversity.'

The example we choose here is coral reefs. Corals are very small, colonial animals that live in close association with tiny algae in a partnership that benefits both organisms. Corals provide shelter and waste products for the algae, while the algae provide sugars and nutrients that corals need to survive (Pockley, 2000). Corals are sensitive to sea temperatures: a temporary rise of only one or two degrees above the normal seasonal maximum can cause 'bleaching', the ejection of algae from their tissues (Figure 6.22). Corals can usually recover, if the temperature changes are short term, as long as they are in good health. Unfortunately, in many areas this is not the case, as they are under threat from a variety of human activities including pollution, coastal erosion, fishery activity and tourism.

The record breaking El Niño event of 1997 and 1998 allows us to preview the climate threat to coral ecosystems. It was responsible for unusually high sea

Figure 6.22 Coral reefs (a) healthy and (b) bleached.

surface temperatures over many tropical regions, particularly in the Indian Ocean and the Caribbean, and led to the most widespread 'bleaching' of corals ever recorded. Just before the El Niño event, a study to investigate the human impact on corals found 11 per cent of the world's reefs destroyed. Two years after the bleaching event a second study from the Global Coral Reef Monitoring Report found 27 per cent of the global reefs effectively lost, and a further 30 per cent on the 'critical' or 'threatened' list. According to the report the losses were heaviest in the Indian Ocean where temperature anomalies were highest. Here, 59 per cent of coral reefs were lost. However, there was a reasonable chance of recovery for remote reefs (i.e. those least damaged by human activity). It is the impact of multiple stresses on ecological systems that does the damage. Corals already under stress are less likely to recover from the bleaching, and while this strong El Niño event is not the same as global warming it is a foretaste of what might happen more frequently in a warmer world. To help us appreciate the full spectrum of threats that climate change poses for vulnerable communities and systems, our attention will now focus on extremes of climate change.

○   Describe (a) the sensitivity and (b) the vulnerability of coral in the
     Indian ocean to raised sea surface temperatures.

●   (a) Corals are very sensitive to maximum sea surface temperatures.
     A rise of one or two degrees celsius can cause bleaching. (b) Their
     vulnerability depends on their condition before the event (and its
     duration). Healthy corals usually recover quickly, but those already
     stressed or damaged may not, they are much more vulnerable.

5.2 Extreme climate events and vulnerability

The climate hazards to human society from extreme events include the damage, hardship and death caused by droughts, floods, heat waves, windstorms and avalanches. Our example here is the impact of some recent floods.

The reinsurance industry is the payer of last resort for major disasters and has good reason to monitor the human and economic costs arising from 'natural' disasters such as major storms, floods and earthquakes. In recent decades there has been a sharp increase in the global impact of major weather or climate related disasters (Figure 6.23).

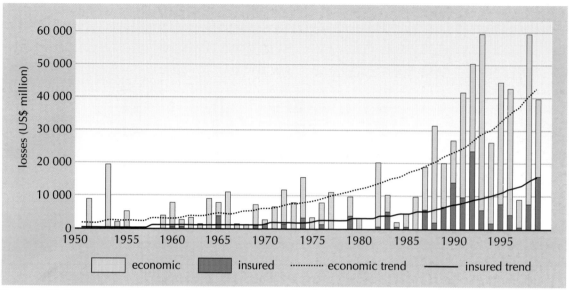

Figure 6.23 The economic impact of major weather and flood disasters from 1950–1999. *Source*: adapted from IPCC, 2001d, Figure 8-1.

Graphs such as these have sometimes been used to suggest that an increase in climate variability or extreme weather has taken place in recent decades.

○   Why should you be wary of such a claim?

●   The extent of impacts from disasters, in this case economic and insured losses (which are themselves estimates), depend on many factors in addition to the severity of weather and climate, in particular the increased numbers of people living in areas at risk such as coastal zones, and the higher values of properties. It would appear that in many areas economic development is leading societies to become more sensitive to extreme 'natural' events, but this is occurring in different ways, and some are more able to cope or adapt (i.e. are less vulnerable) than others.

We can illustrate this latter point by looking at what happened in a single year, 2000, using data from one of the world's major insurers of last resort, Munich Re. According to them, it was a comparatively quiet year, marked by floods, but with few major storms or earthquakes.

Table 6.6 compares the losses from natural disasters on two continents, Africa and Europe, one poor, one rich, for the year 2000. A clear pattern is apparent: economic and insured losses are far greater in rich Europe, largely because the buildings that are destroyed or damaged have a higher value. This doesn't mean that destruction is less in Africa, only that the value of the buildings damaged or destroyed is less. In contrast, loss of life in Africa is greater than in Europe because the infrastructure to protect and save people is less developed. These disparities are typical of rich and poor countries worldwide.

Table 6.6   Losses from natural catastrophes, 2000

Region	Africa	Europe	Global
Number of loss events	66	178	851
Economic loss (US$ m)	690	13,160	31,060
Insured loss(US$ m)	60	2,150	8,270
Number of fatalities	1,490	410	9,270

Note: Weather and climate-related events account for nearly 80-90 percent of the total of each figure, apart from fatalities in Europe, which make up a lower proportion and are overstated here.

The comparative vulnerabilities of Europe and Africa can be illustrated by comparing two major flood events in 2000. In the autumn, record levels of rainfall across much of Europe caused widespread floods, of which two were notable. In October, flash floods, mudflows and landslides in the Swiss and Italian Alps badly damaged mountain villages and flooded the northern plains of Italy, claiming 38 lives and generating losses of approximately US$8.5 bn. In the UK the same weather patterns led to the wettest autumn in the 250 year record for England and Wales, leading to widespread river flooding. Property damage was estimated at US$1.5 bn and 10 lives were lost. In the UK the Environment Agency reported that 10 per cent of the population and 1.8 million homes 'are now located in potentially flood prone areas' due to a combination of projected climate change

Figure 6.24 The Mozambique village of Chinhacanine in February 2000, showing settlements flooded to the roof tops after heavy rains and a cyclone.

and the practice of the last 50 years of building homes and towns on river flood plains (Environment Agency, 2002).

In south east Africa, heavy rains caused extensive spring flooding to Mozambique and surrounding areas that lasted for several weeks (Figure 6.24). Half a million people were made homeless and altogether five million, about a quarter of the population, were affected. This single event accounted for more than half the economic and population losses in Africa for the year. Although many individuals and communities have been deeply affected by the floods in the Alps and the UK, the worst effects were at local, not national level. In contrast, the economy of Mozambique has been devastated. In most cases poorer countries are much more vulnerable to extreme events than richer countries because their existing infrastructure is weaker and they have less capital to renew it. (See Chapter 7 for more on uneven vulnerability).

5.3 Hazards from future large-scale discontinuities

A different form of extreme climate change and of risk type is presented by the possibility of large-scale, irreversible changes. Human-induced climate change has the potential to trigger major changes in Earth systems that could have severe consequences at regional or global scales. Concerns have been expressed about two particular events being triggered by global warming.

One is the sudden disintegration of the West Antarctic ice sheet, which would cause a sudden global rise in sea level of six metres if it occurred. There is concern that this massive ice sheet might be unstable, in which case once it is released there would be little to stop its collapse. Glaciologists think that this eventuality is 'very unlikely' during the present century but on longer timescales are less sure.

The other concern is for the ocean thermohaline circulation (THC) in the North Atlantic region (see Chapter Five). At the moment the surface circulation transports warm water to high latitudes in the North Atlantic and is responsible for the mild winters of Western Europe. Most climate models show this circulation weakening with global warming, which would reduce heat transport to the Atlantic and Europe, but none projects a net cooling because the effect of greenhouse warming still predominates. There is concern that the THC might suddenly 'switch off' in the space of a decade or so, because Greenland ice core records and ocean sediments show that this has happened many times during glacial periods. However, the likely cause of changes in the past – sudden surges of ice into the Atlantic from North American ice sheets – are quite different from the mechanisms suspected under global warming. Much more likely is the partial shutdown of the THC, a much less threatening scenario but one still likely to have major regional impacts.

These are both examples of possible outcomes that could have far reaching impacts and could both be irreversible. Their likelihoods are not known, they are thought to be very low, but could increase significantly if climate change accelerated.

5.4 The impacts of cumulative climate change

In this final example we try to imagine what parts of the UK would be like towards the end of this century, but with the aim of building a cumulative picture rather than an economic balance sheet.

Key climate projections for Europe are of milder, wetter and stormier winters particularly in the north, and hotter and drier summers particularly in the south. Some of the key impacts come from extreme events, both short term and seasonal (Table 6.7).

Table 6.7 Climate sensitivities in Europe

Sensitivity	Comment
Extreme seasons	
Mild winters	Increased likelihood of pests and potential spread of some disease vectors (e.g. mosquitoes).
Exceptionally hot, dry summers	Major effects on agriculture and water supply and natural environments. Increased pollution levels (summer smog) and associated health risks, including heat stress.
Short duration hazards	
Windstorm and possible tidal surges	Severe windstorms have affected western Europe in 1987, 1990 and 1999, causing extensive damage.
Heavy rain leading to river-valley flooding and flash floods	Severe flooding occurred across Europe in 2002.
Slow long-term change	
Coastal loss and erosion	Coastal habitats are squeezed between rising sea level and sea defences.

Europe is less vulnerable than many poorer regions, but still has vulnerable sub-regions, particularly in some of its less wealthy and more marginal areas: the south and the European Arctic. Its remaining natural ecosystems are also vulnerable from shifting biotic zones. Some species would be threatened by the loss of wetland, tundra and isolated habitats from a combination of climate change and human pressure.

The UK occupies a more favourable situation. It also straddles the trend of future climate to a wetter north and drier south. Scotland is expected to become wetter, particularly in winter, as well as warmer, while in the south, wetter winters are accompanied by hotter, drier summers.

One effect of a warming climate is the northward shift of natural biomes, A projected temperature rise of between 2 °C to 4 °C during the current century would lead to an estimated average movement per decade of 50 to 100 km. To explore what this might mean, imagine instead moving a city like London *south* to a warmer climate.

A 2 °C rise would give it a climate similar to Bordeaux in the south west of France, with a similar mean and seasonal range of temperatures but, surprisingly, much less summer rainfall – agriculture will not be as productive as it currently is in the Bordeaux area. At the higher end of the scale a 4 °C rise would take London into the Mediterranean. Marseilles, for example, gives a good match for both temperature and precipitation. Of course other elements of the weather would be different: wind and sunshine may not quite compare so favourably and the Thames Riviera won't be quite the same as the Cote d'Azur. Our exposure to climate extremes may also be less than welcome; here are two recent predictions for the 2080s:

- Extreme summer heat and drought, expected only once or twice in a century, e.g. an August as hot as that observed in 1995 – the second hottest August on record – occurs on average every other year by 2080 (Hulme et al., 2002).
- Extremely wet winters like that of 2000 will become five times more frequent across northern Europe and can be expected once every eight years in the coming century (Palmer and Raisanen, 2002, pp.512-4).

What would this change mean for future generations living in southern England? Consider for a moment just two aspects of the environment (housing and landscape) in the following activity.

Activity 6.7

What might be the effects on housing and landscape if the climate of London became like the climate of Marseilles?

Comment

Most UK housing stock and urban planning is built for a cool, windy climate not to cope with summer heat. A considerable proportion might have to be redesigned to reduce summer heat stress for the population, and the buildings themselves protected from twin threats of flooding and drought-induced subsidence. The 10 per cent of buildings in flood risk areas would need considerable modification, protection and may be abandoned. The effect on the typically English landscape would also be dramatic. Agriculture would need to adapt to something more appropriate to a Mediterranean type of climate, but other elements of the landscape would not be able to keep pace with the rapid shifts as a temperate woodland pattern gives way to the vegetation zone of Mediterranean scrub. The English landscapes celebrated by artists would disappear.

○   At some stage in our imagined journey in a warming world we
      crossed an important climate threshold. What is it?

●   We moved from a temperate westerly to a Mediterranean climate,
      and from the zone of temperate forest to Mediterranean scrub-
      land. See Chapter 1 for a discussion of global vegetation patterns.

Activity 6.7 illustrates that with rising temperatures, not only is the frequency of
potentially adverse events increased, but there is also the chance that a climate
threshold will be crossed, which can lead to major qualitative changes in our
environment. Figure 6.25 takes up this point, and applies it to five categories of
impact from global climate change. It shows graphically how hazards change
both qualitatively and quantitatively with increasing global mean temperature.
We have discussed examples from three of the categories (I, II and V) here, the
other two have been touched upon but belong more to the social and economic
sphere which is taken up in the next chapter.

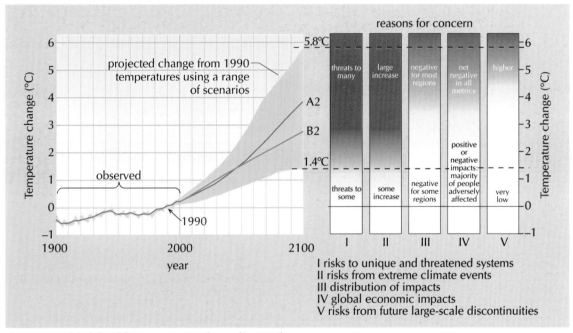

Figure 6.25 Models of future concerns from climate change.
Source: adapted from IPCC, 2001d, Figure SPM-2.

5.5 The risk of global climate change increases

In Sections 5.2 to 5.4 we have introduced four different types of climate change
impact, each with its own distinctive set of hazards. In many of the examples the
chances of adverse effects appear to increase with the rise of global temperatures,
but not always at an even pace and often distributed unevenly. In other examples,
major new hazards may appear as some threshold in the climate system is crossed.
Their likelihood of occurrence may be low but increase as global climate changes.

Figure 6.25 attempts to bring together this varied calculus of risk, to show how the likelihood of adverse effects might vary as global temperatures rise by up to six degrees, a range that encompasses the projections of the IPCC Third Assessment Report for the end of this century: 1.4 ° to 5.8 °C. Although the details remain hazy, clearly somewhere along this temperature scale the threshold introduced at the start of this section is crossed: dangerous interference with the climate system.

Summary

The effects of climate change are both widespread and diverse. Certain natural systems, such as coral reefs, are particularly vulnerable, particularly if already stressed. Human societies also suffer from extreme weather, including storms and floods, and the poor are the more vulnerable. As global climate change proceeds, biomes found within particular areas will change, with associated risks to humans and the ecosystems in which they live, and, more generally, the likelihood of change to the Earth's biological and physical systems increases.

6   Conclusion

As a free resource we tend to take our atmosphere and climate for granted, but they are highly susceptible to change. A range of effects arises from simply altering the concentrations of many minor components, both gases and aerosols, or by introducing new ones. This has long given rise to local air pollution, and more recently to regional pollution. It has now become a global issue known as global climate change. In exploring changes to the global climate system, we have seen that the atmosphere is deeply integrated with the other components of the Earth system. A change to one is change to all, and an injury to one is, usually, an injury to all.

References

Bingham, N. and Blackmore, R. (2003) 'What to do? How risk and uncertainty affect environmental responses' in Hinchliffe, S.J. et al. (eds).

Blowers, A.T. and Smith, S.G. (2003) 'Introducing environmental issues: the environment of an estuary' in Hinchliffe, S.J. et al. (eds).

Environment Agency (2002) 'Averting flood disaster', http://www.environment-agency. gov.uk.our_work/flooding/342401/?lang=_e&region=&projectstatus=&theme= &subject=&searchfor=averting&topic=&area=&month=, accessed 12 December 2002.

Hinchliffe, S.J., Blowers, A.T. and Freeland, J.R. (2003) *Understanding Environmental Issues*, Chichester, John Wiley & Sons/The Open University (Book 1 in this series).

Hulme, M., Jenkins, G.J. and Turnpenny, J.R. (2002) *Climate Change Scenarios for the United Kingdom: The UKCIP02 Briefing Report*, Norwich, Tyndall Centre, University of East Anglia.

Intergovernmental Panel on Climate Change (IPCC) (2001a) *Climate Change 2001: Synthesis Report. Third Annual Report of the IPPC*, UNEP/WHO, Cambridge University Press.

Intergovernmental Panel on Climate Change (IPCC) (2001b) *Climate Change 2001: The Scientific Basis. Third Assessment Report of Working Group I*, UNEP/WHO, Cambridge University Press.

Intergovernmental Panel on Climate Change (IPCC) (2001c) *Emissions Scenarios: Mitigation. Third Assessment Report of Working Group II*, UNEP/WHO, Cambridge University Press.

Intergovernmental Panel on Climate Change (IPCC) (2001d) *Climate Change 2001: Impacts, Adaptation and Vulnerability. Third Assessment Report of Working Group II*, UNEP/WHO, Cambridge University Press.

Palmer, N.T. and Raisanen, J. (2002) 'Quantifying the risk of extreme precipitation events in a changing climate', *Nature*, vol.415.

Penuelas, J. and Filella, I. (2001) 'Responses to a warming world', *Science*, vol.294.

Petit, J.R., Jouzel, J., Raynaud, D., Barkov, N.I., Barnola, J.M., Basile, I., Benders, M., Chappellaz, J., Davis, M., Delayque, G., Delmotte, M., Kotlyakov, V.M., LeGrand, M. Lipenkov, V.Y., Loritius, C., Pépin, L., Ritz, C., Saltzman, E. and Stievenard, M. (1999) 'Climate and atmospheric history of the past 420,000 years from the Vostok ice core in Antarctica', *Nature*, vol.399.

Pockley, P. (2000) 'Global warming identified as main threat to coral reefs', *Nature*, vol.407, p.932.

Stokstad, E. (2001) 'Myriad ways to reconstruct past climates', *Science*, vol.292, pp.658–9.

World Health Organization (WHO) (1980) *Glossary on Air Pollution*, Copenhagen, WHO.

Answers to Activities

Activity 6.1

Approximately 6 km. The difference in temperature is 15 °C - (-19 °C) = 34 °C. If the lapse rate is 6 °C per km, this gives an altitude of 5.7 km.

Activity 6.2

Table 6.3 A comparison of climate zones and climax vegetation types

Climate zone	Climax vegetation classification
inner tropical zone	tropical rainforest
outer tropical zone	savannah
sub-tropical dry zone	desert
sub-tropical winter rain	steppe (and Mediterranean shrubland)
moist temperate zone	temperate deciduous forest
boreal zone	boreal coniferous forest
sub-polar zone	tundra

Activity 6.3

a) The observed increase in the twentieth century of approximately 0.6 °C is not exceptional from the perspective of the last 10,000 years, which has experienced several global variations of 1 to 2 °C.

b) The forecast of a rise of global mean surface temperature between 1.4 °C to 5.8 °C is highly significant. A rise in global temperature of more than 2 °C is likely to be greater than anything experienced during this period, and the higher end of the range (4 °C to 5.8 °C) approaches the size of the changes seen only during the transition from glacial to interglacial conditions.

Activity 6.4

Water vapour feedback enhances the initial change of a warming in the atmosphere, so it is an example of positive feedback (Section 4.2). Several stages can be identified in the feedback loop. A warmer atmosphere and surface increases evaporation leading to more water vapour in the atmosphere. This increases the greenhouse effect which, in turn, leads to more warming of the atmosphere. These stages form the components of the

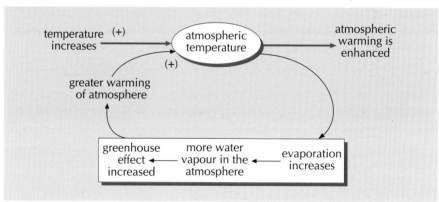

schematic diagram (see also Figure 6.15).

Activity 6.5

a) The balance would change little for many decades because carbon dioxide concentrations would persist, as would global warming.

b) Sulphate aerosols would disappear from the atmosphere within a few weeks and with it their negative influence on the radiation balance.

Activity 6.6

According to Figure 6.20, the contribution to recent global warming from the well-mixed greenhouse gases is approximately 2.4 ± 0.2 W/m², more if the effect of tropospheric ozone is included. The estimate for the solar contribution lies between 0.1-0.5 W/m², although this is poorly understood. Based on what we know, it appears that greenhouse gases have had more of an effect than the sun on recent global warming.

Chapter Seven

Uneven development, globalization and environmental change

Noel Castree

Contents

1   Introduction

By this point in the book you will have come to understand that environmental changes are constantly occurring. Even where human activities have little or no environmental effect, natural systems are rarely static. When humans do exert their influence upon the land, water bodies or the atmosphere – and there are now over 6 billion of us on the planet doing just this – the result is often a slowing down, speeding up or radical transformation of natural processes. This is why understanding environmental changes requires us to consider both natural and human factors. We need to comprehend not only the physical processes causing such change but also the human actions that trigger, halt, accelerate or even generate such processes. In other words, much environmental change, both past and present, is the result of a **'human–environment dialectic'**: that is, humans alter environments (and not always intentionally or knowingly) and these altered environments, in turn, may affect human behaviour and well-being.

Different societies have been adapting to and altering their local environments for millennia, as you've seen in the previous chapters. This is one reason why the world has been and remains so environmentally diverse. However, two things are arguably new and distinctive about human–environment relationships in the contemporary world. First, these relations are increasingly *international* or *global* in scale. That is, local human activities can have global environmental effects while those effects, in turn, can influence different local populations and local environments. Second, the more globalized or 'stretched' human–environment relations have become, the more *spatially and temporally uneven* their causes and consequences seem to be. That is, different societies contribute to, and are affected by, global environmental changes to different degrees, in different ways, and at different times.

We can briefly illustrate these two points with an example. Look at Figure 7.1, a newspaper article on a topic you will certainly have heard of and have studied in Chapter Six: global warming. This is caused by the accelerated release of certain gases, such as carbon dioxide, into the atmosphere by different human communities in different parts of the world. As you read in the previous chapter, these gases can cause radiative forcing and contribute to an enhanced natural 'greenhouse effect'. As the article intimates, if temperatures rise by a degree or more in the decades ahead, then the environmental impacts worldwide will probably be profound. So lots of *separate*, local-scale carbon dioxide emissions *together* help to produce a worldwide environmental change (global warming).

However, different societies contribute to, and experience the effects of, global warming differently. For instance, although it contains only 4 per cent of the world's population, the USA is currently responsible for 12 per cent of total annual emissions of carbon dioxide. While the USA may suffer harm from the environmental changes likely to arise from greenhouse warming, it is poor developing countries that will probably be the worst affected. Take

human–environment dialectic

Greenhouse build-up worst for 20m years

Tim Radford
Science editor

Carbon dioxide levels in the atmosphere are at their highest for 20m years, two British scientists reveal today.

The two men analysed the shells of tiny ocean creatures to build up a fossil record of the chemistry of the atmosphere for the past 60m years. Their discovery will add to pressure on governments to implement international agreements to reduce the greenhouse gases that fuel global warming. The news comes hard on a report from the University of Colorado that Arctic temperatures are at their warmest for at least 400 years.

Paul Pearson of the University of Bristol and Martin Palmer of Imperial College, London, report in Nature today that they used the evidence of plankton shells drilled from the seabed to estimate the acidity of the sea water over a span of time back almost to the era of the dinosaurs. Carbon dioxide dissolves in water to form a weak carbonic acid. They reasoned that the tissue of floating organisms would reflect the carbon dioxide levels of the world around them – and then hold that record locked in the fossilised shells.

Other studies of more recent records have confirmed a picture of a rapid warming in the past two decades. Winter is in retreat. The growing season in Europe is 11 days longer than it was 35 years ago. Sea levels have crept higher throughout the century. Six of the 10 warmest years ever recorded occurred in the 1990s; the other four all happened in the late 1980s. The Arctic ice cover is in retreat, shrinking by an area the size of the Netherlands every year. It is also thinning, from more than three metres thick to less than two metres in 30 years.

Now a team from Boulder, Colorado, reports this week in the Dutch journal Climate Change that wherever it looked – at Arctic ice cores, lake sediments or plant growth in permafrost – it saw confirmation that the Arctic had warmed by 6°C in the past 30 years.

But in the tropics, humans have felt the impact more cruelly. Researchers long ago predicted more storms, droughts and floods as a consequence of a warmer world, and both the economic costs, and the numbers of victims of climate-related disasters, have increased in the past decade.

Paul Epstein, of the Harvard medical school in the US, warns in this month's Scientific American that global warming must also mean an increase in water-borne and insect-borne diseases to add to the threat.

But until now researchers have had no clear idea of the long term pattern of the planet's atmosphere. Dr Pearson said: "Our observations put the modern greenhouse effect into a long term perspective."

For much of the past 20m years, the planet has been quite cool. An Australian team reports in Nature today that at the height of the last ice age, 21,000 years ago, sea levels were 135 metres lower than today, and the continents were covered by an extra 52m cubic kilometres of ice.

But carbon dioxide levels are rising swiftly because of fossil fuel burning and the clearing of the planet's forests, which in past aeons have taken carbon from the air and stored it, first as wood and then as coal. By 2100 the carbon dioxide levels will increase to match those last seen in the Eocene, 50m years ago. In those days, much of Europe was flooded, there were no ice caps, London was a steaming mangrove swamp and the average temperature of southern England was 25°C. Today, the average temperature is 10°C.

Professor Palmer said: "This does not necessarily mean we will recreate Eocene-type conditions. There are still too many unknowns involved in climate prediction. But the sweltering ice-free world of the Eocene does warn us of what might happen if a runaway greenhouse effect sets in."

Figure 7.1   *Source: The Guardian*, 17 August 2000.

Bangladesh, an economically poor Asian country whose population is just 50 per cent that of the USA. It currently emits just 0.4 per cent of world atmospheric carbon dioxide. But despite being a minor atmospheric polluter, it is likely to suffer disproportionately from the environmental effects of continued global warming. For example, if world sea levels rise because of thermal expansion of the oceans (water volumes increase with heat), then as much as one-third of Bangladesh's land area may be submerged over time. This inundation by water will destroy existing land-based ecosystems in the country and gradually displace tens of millions of people. It will, in other words, be a major environmental and social problem for this already poor country. In turn, and thinking about the future, because Bangladesh's economic growth will perhaps be slowed down, the country's already small carbon dioxide emissions will probably remain quite static. Bangladesh will thus, in the years ahead, be even *less* responsible for the global warming potentially causing it so much environmental and social havoc!

In this chapter we shall explore in some detail how globalization, uneven development and environmental changes are related. Specifically, we will develop an answer to two questions:

1   What does it mean to say that the relationships between humans and their environments are increasingly 'globalized'?

2   What are the causes and consequences of these changes across space and through time?

As my brief discussion of global warming indicates, these are very important questions that, in the most extreme cases, relate to matters of life or death. As we will see, answering these questions gives us key insights into the human dimensions of environmental changes in the contemporary world. These insights, as we will also see, entail understanding how space and time are of fundamental importance in explaining environmental changes. Put simply, human actions in one place and at one point in time can have environmental effects on people in another place at the same time, or at a later time. Consequently, developing answers to our two questions entails (a) figuring out *who is responsible* for environmental changes in a highly interconnected world and (b) determining the *significance* of such changes for different people in different parts of the world at different times. However, as we will also see, there is disagreement about what the correct answers to our two above-stated questions are. This is because it is possible to explain the human causes and consequences of environmental change in more than one way.

As you discovered in Chapter Three, one popular model that social scientists have used to explain the human dimensions of environmental changes is the Malthusian model. In this chapter I shall introduce two different approaches: these are the 'tragedy of the commons' and 'livelihoods' models. These models, as I will explain in due course, have much to say about how, and with what effects, environmental changes are connected to globalization and uneven development. We will compare and contrast these models in order to highlight which, if either, offers a better explanation of environmental changes in an interconnected world. So, in effect, if we are to properly answer the two questions being asked in this chapter, we also have to pose a third, namely:

3   How should we model (that is, represent and explain) the links between environmental changes, uneven development and globalization?

We shall begin, in the next two sections, with some definitions of globalization and uneven development where we see how both relate to environmental changes. In Section 4 I build on this by challenging the idea that the globalization of human–environment relationships has only occurred in the last few years. These three sections together provide a set of *descriptive* or *analytical* concepts for understanding globalization, uneven development and environmental changes.

In the remaining sections of the chapter we seek to *explain* the causes and consequences of these changes. In Section 5 I will compare and contrast the tragedy of the commons and livelihoods explanations of environmental change. This leads, in the penultimate section, to a more detailed exploration of the livelihoods model. In conclusion, I reflect upon future environmental changes in a world of spatial and temporal inequality.

2   'Globalizing' environmental changes ...

2.1   What's 'global' about contemporary environmental changes?

Global warming is just one of several new global environmental changes caused by humans (so-called '**anthropogenic**' changes). Other well-known changes of this kind include the thinning of the stratospheric ozone layer (among other things stratospheric ozone protects us from the harmful effects of ultraviolet radiation from the sun) and 'acid rain' (that is, the deposition on land and in water of nitrous and sulphuric pollutants many kilometres from their point of origin). In the early twenty-first century, it is frequently said, many environmental changes have become globalized. But what does this mean? It does not mean the same thing as the term '**globalization**', as we normally understand it, which conjures up images of increasing socio-economic homogeneity spearheaded by the likes of McDonald's, Coca-Cola and Nike. Instead, it refers to the propensity of human actions to have international and increasingly global-scale environmental impacts.

anthropogenic

globalization

As the case of global warming suggests, we are talking here about 'unnatural' changes resulting from human activities. Though humans have been altering environments for generations, one of the unique things about the modern world is the *spatial scale* of alteration.

Activity 7.1

With this in mind, ask yourself: what is 'global' about an environmental change such as global warming?

Comment

The simplest answer is that this warming is global in the sense that it is not limited to any one part or region of the world's atmosphere. But a more precise and complex answer is that *humans worldwide are both causing and being affected by this phenomenon*. To be more specific:

- individuals, communities and industries in every country on the planet are emitting gases, such as carbon dioxide, that are causing global warming;
- equally, because this warming affects the entire atmosphere, it will ultimately affect every different country and community on the planet in some way, shape or form.

Phenomena such as global warming and ozone layer thinning are what we might call 'global environmental changes with local causes and local consequences': see Figure 7.2. That is, they both arise from and affect the actions of different people in different places using different environments around the world. The word 'global' refers here to a mixture of international and truly globe-girdling environmental changes. These changes involve events at one spatial scale – the local or regional – having knock-on effects at another scale – the international or global (and vice versa). There is, however, a second category of global environmental changes. To understand this second category, let us consider another example: namely, the logging of the Amazon rainforest in Brazil. The Amazon rainforest is one of the largest remaining areas of tropical hardwoods on Earth (it covers an area roughly the size of western Europe). In recent years, commercial logging companies have harvested these trees and exported them to various other countries around the world, including the UK. Once abroad, the wood is made into consumer products such as furniture. It has been estimated that an area of Amazonian forest the size of Wales is cut down each year and environmentalists now worry that the rainforest may soon

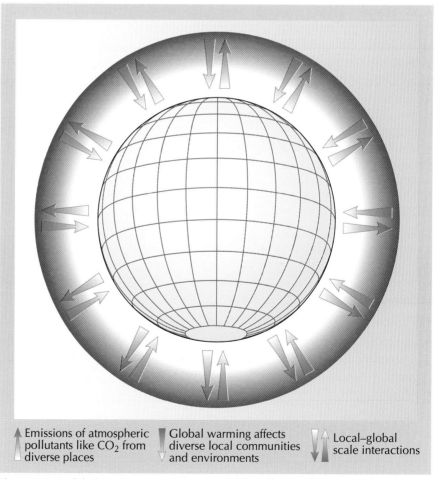

Emissions of atmospheric pollutants like CO_2 from diverse places

Global warming affects diverse local communities and environments

Local–global scale interactions

Figure 7.2 A global environmental change with local/regional causes and consequences: global warming.

Figure 7.3 Deforestation in the Amazon basin.

disappear altogether (Chapter Four, and Greenpeace, 2001): see Figure 7.3. Though commercial foresters are by no means the only cause of deforestation, they do play an important role.

Activity 7.2

Can you figure out what is 'global' about the destruction of Amazonian forest?

Comment

This may seem a strange question because the Amazon rainforest is clearly not a global environment. It is, rather, a local or regional environment that is specific to Brazil and its neighbouring countries. So perhaps the question to ask yourself is this: in what ways is a local- or regional-scale environmental change (in this case Amazonian deforestation) *also* global in its causes and consequences? What answers have you come up with? I can think of at least two:

- One reason that logging companies are currently cutting down so much of the Amazon is to sell wood to global markets. Countries worldwide buy timber from Brazil and the demand for it remains high.
- Trees naturally 'lock up' greenhouse gases such as carbon dioxide. The destruction of Amazonian rainforest, therefore, just might exacerbate global warming by removing an important natural 'sink' for these gases.

Seen in these two ways, Amazonian deforestation might be called a 'local/regional environmental change with global causes and consequences' (see spatial-scale switching

Figure 7.4). As with the first category mentioned above, a change like this results from **'spatial-scale switching'**. That is, global events have local environmental impacts and local environmental changes have global consequences. Amazonian deforestation is among the most well-known and dramatic examples of this second type of global environmental change. But there are many other equally serious, if less visible, changes of this kind occurring all around us (and we will consider some of these later in the chapter).

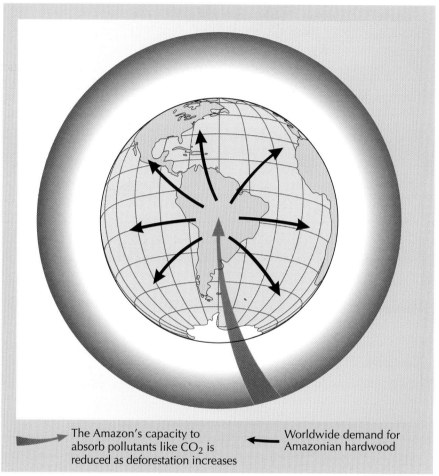

The Amazon's capacity to absorb pollutants like CO_2 is reduced as deforestation increases

Worldwide demand for Amazonian hardwood

Figure 7.4 A local/regional environmental change with global causes and global consequences: Amazonian deforestation.

2.2 Two types of globalization

Thus far we have identified two ways in which environmental changes have become 'globalized', the common denominators being (a) that these changes have human causes and (b) they involve spatial-scale switching (local events affecting global events and vice versa). In reality, our two types of global environmental change are not always easy to distinguish and might best be seen as two perspectives on the same thing, that is anthropogenic environmental changes that possess *simultaneously* local and translocal dimensions. But there is also a further important distinction to make. Let us look again at greenhouse warming and Amazonian deforestation to help us understand what this distinction is and why it matters.

In both cases we saw, in rather different ways, how the environmental changes were either global in scale or had global impacts. This is what we might call *the globalization of environmental change*. However, in both cases we also saw a rather different process at work: that is, *globalization as a cause of environmental change*. The two processes are not quite the same, even though I've used the same word – 'globalization' – to describe them both. When we talk about the globalization *of* environmental change we are simply referring to the large spatial scale at which human activities are now having an environmental impact. But when we talk about globalization as *a cause of* environmental change we are referring to the scale of these *activities themselves*, and how their global organization is having environmental impacts across the whole face of the Earth.

As the geographer John Allen (1995, p.107) has put it, in this latter sense, 'Globalization ... refers to the fact that people in various parts of the world, which hitherto may have been largely unaffected by what happened elsewhere, now find themselves drawn into the same **social space** ...'. Note the use of the term 'social space' in this definition. What Allen is saying is that in the modern era the connections between people are no longer predominantly direct, intentional, local, face-to-face connections – as they generally were in the pre-modern world. Rather, they are increasingly global connections that are 'stretched' across the face of the Earth. These stretched connections may be direct and intentional or indirect and unintentional (see Figure 7.5).

social space

Activity 7.3

Think again about the two examples we have been discussing. In what senses is globalization a cause of these environmental changes? If you think hard about this question, you should come up with quite a few answers.

Comment

In the case of Amazonian deforestation the answer is fairly obvious. It is the trade connections that link Brazil with foreign markets that are largely driving the harvesting of tropical hardwoods. But what about atmospheric warming? The answer here is less readily apparent because its causes seem to have little to do with direct, intentional global connections between different peoples. They appear, instead, to have more to do with different peoples *independently* polluting the atmosphere in different places and thereby *unintentionally* affecting peoples many miles away (for instance, in Bangladesh). However, to anticipate our later discussion of the livelihoods model a little bit, another way in which globalization has helped to cause atmospheric warming is related to the global spread of western notions of economic development. Collectively, so-called 'less developed countries', like Bangladesh, emit some 63 per cent of annual greenhouse gases (IPCC, 2001). Though some observers (like Malthusians) see this as primarily a function of high population levels, it is also arguably a result of the *form* of economic development. Many developing countries are currently following the path paved by the developed world, even though this is not necessarily the best or most appropriate path to follow. The long history of connections between western and non-western countries has led to the spread of many western values worldwide. Some of these values relate to development. That is, many non-western nations are trying to develop their economies on the basis of heavy industries, like coal-mining and steel-making, and mass transportation (like cars) – just as western countries did during the nineteenth and twentieth centuries. These industries and means of transportation are highly polluting and among the principal culprits for the increase in greenhouse gas emissions in recent decades.

Identifying these two different meanings of globalization in the context of environmental change is important. We now start to understand why explaining global environmental changes, of the two kinds identified in this section, involves an appreciation of the global scale of human relationships, be they relationships of trade, culture, or what have you. In the modern world, then, environmental changes of all kinds are increasingly the product of the global ties linking distant peoples in different places. However, this said, not all environmental changes

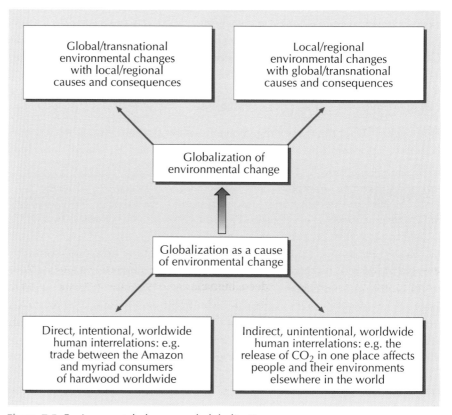

Figure 7.5 Environmental change and globalization.

today, and certainly not in the distant past, can be attributed to the globalization of human relationships. As you've seen in several of the previous chapters in this book, it remains the case even today that many environmental changes, including natural ones, are essentially local or regional in their causes and effects.

Summary

- In answer to the first main question posed in this chapter there are four senses in which we can talk about the 'globalization of environmental changes'.

- Global environmental changes are of two related kinds, 'global environmental changes with local causes and consequences' and a 'local environmental change with global causes and consequences'.

- Globalization means two different but related things in the context of these global environmental changes. It refers to the global scale of those changes themselves and, secondly, the global scale of the human activities causing those changes.

- Globalization as a cause of environmental changes can involve direct and intended or indirect and unintended human impacts.

3   … in an uneven world

Having explored what is meant by globalization in the context of contemporary environmental changes, let us turn to the second question I posed at the outset: how is **uneven development** both a cause and a consequence of environmental changes in a globalizing world? To answer this question – and to understand why it is even worth asking – we will obviously have to explore what is meant by 'uneven development'. But in order to do that, it is first necessary to explore the concept of *environmental differences*.

uneven development

3.1   Environmental differences

Environmental differences are both natural and anthropogenic. Geology, climate and evolution have provided an astonishing array of land, air and aquatic environments. These environments vary spatially and temporally, from the tundra of Siberia to the rainforests of Costa Rica. Since we came into existence as a species, humans have adapted to or altered these different environments in order to survive and prosper. Indeed, humans and environments would probably perish without environmental diversity. Humans require this diversity to meet their nutritional and other needs, while non-human organisms require myriad other organic and inorganic parts of nature in order to survive. Even today, when so many environmental changes have been wrought by humans, environmental differences remain – it's just that many of the differences are now anthropogenic rather than simply natural. For instance, despite the immediate, if long-distance, connections between a Brazilian logger and a well-to-do Londoner buying a mahogany (a tropical hardwood) table from an expensive furniture dealer, both

Figure 7.6 (a) High-class English furniture shop: demand for mahogany furniture remains high, but most tropical hardwoods now come from sustainable sources.

Figure 7.6 (b) A mahogany seedling in Brazil: it will take at least 25 years before it is ready to be logged.

individuals live in, and draw upon, very different biophysical environments for their survival and well-being, although, clearly, part of the Londoner's 'environment' extends all the way to the Amazon: this is an example of the 'ecological footprint' idea introduced in Chapter Four (see Figure 7.6). So the globalization of environmental change and human activities – as discussed in the previous section – has *altered*, rather than simply *undermined*, environmental differences worldwide. This said, many ecologists do believe that environmental diversity has diminished in the last few decades because of human actions. Nonetheless, this belief does not imply that environmental differences have disappeared but, rather, they have been reduced.

Activity 7.4

Let us explore these connections between environmental differences and globalization a little further. Returning to one of our earlier distinctions, can you identify how environmental differences help to cause the two types of globalization identified? Think again about global warming and the logging of Amazonia. Try to pinpoint the reasons why environmental differences lead to our two types of globalization.

Comment

How many answers did you come up with? Possibly quite a few. The following may have been among them:

- *Human alterations of various local and regional environments lead to global environmental changes.* For instance, even though coal is not found in all countries worldwide, the burning of it in different places adds carbon dioxide to the entire atmosphere. Different people using coal in different places produces an environmental change that affects them all. So the effects of coal-burning 'leak out' from the different places where coal is found or used.

- *Environmental differences between places encourage the globalization of human activities.* For instance, precisely because it is physically impossible for tropical hardwoods to grow in Britain's cool climate, our well-to-do Londoner *has* to look elsewhere for his/her hardwood furniture. Specifically, for a mahogany table to be bought in London, connections have to be actively forged between Amazonian loggers, shippers who transport wood across the Atlantic and furniture-dealers in England's capital city.

To summarize, environmental differences are *both a precondition and a cause* of globalization (in both of the senses that we have discussed the latter term) when it comes to environmental changes. But we can take this further. Having identified how environmental differences might affect globalization and environmental change, can we identify how globalization influences environmental differences? The answer is yes and we have, in fact, already done so in the first two sections of this chapter. What we have already seen is that the globalization of environmental change and globalization as a cause of environmental change can *alter existing environmental differences or create new ones.* In the case of global warming we have seen how this transcontinental problem might affect Bangladesh's coastal environments in different ways from those of a country such as the USA. In the case of the Amazon, we can see how global trade and consumption patterns are leading to the possible loss of this unique South American ecosystem.

3.2 Uneven development

So, environmental differences matter a great deal when it comes to understanding environmental change and globalization. And it is here, at last, that we come to the issue of uneven development. Uneven development is a complex concept and has been defined in a wide variety of ways. At its simplest it refers to varying levels of human well-being across space and/or through time. This simple definition is, however, ultimately inadequate. It leaves unanswered the complex question of how one defines development in the first place. The !Kung bush-people of southern Africa do not, for example, drive BMW cars, but does this mean they are 'under-developed'? Or, to take a different case, is the UK 'developed' because it now has the technology to genetically modify crops, even though such modification might do untold environmental harm? Clearly different *social values* are involved in determining what one considers 'development' to be (a theme taken up in **Bingham et al., 2003**). Nonetheless, whatever one's definition, it is obvious that both today and in the past the levels of material well-being of different societies vary and have varied considerably. In the worst cases, some populations are unable to meet even the most basic requirements of existence, namely sufficient food and clean water, as was seen in the terrible famines in the

Sudan and Ethiopia in the 1980s and 1990s (we'll explore the reasons for this later in the chapter).

These often stark variations in levels of development between different places are frequently, though not always, linked directly to environmental differences. The next activity asks you to consider the nature and strength of these links.

Activity 7.5

In light of what you have learnt in the previous chapters of this book, list all the ways in which a community's economic and social development can be influenced by the *local* or *regional* environment the community draws upon. You might want to distinguish between 'environmental constraints' and 'environmental opportunities'. Ask yourself: to what extent will the balance of opportunity and constraint *determine* a community's development level?

Comment

How long is your list? You will probably have noted down a wide range of opportunities and constraints – from soils to minerals to weather. Furthermore,

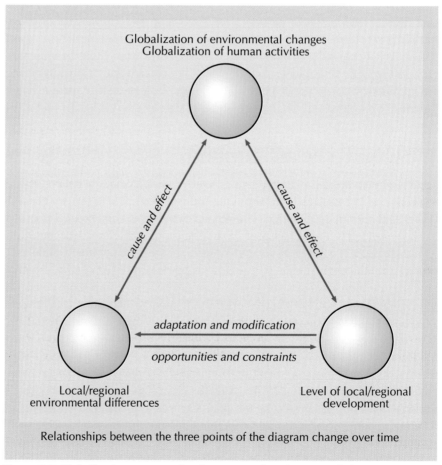

Figure 7.7 Globalization, uneven development and environmental change.

you may have concluded that 'extreme' environments (like arctic or desert environments) correlate with less developed communities, while 'favourable' environments (such as those found in temperate regions) correlate with more developed communities. However, while local/regional environments do indeed present a wide range of opportunities and constraints to those living in them, the correlations are in fact weak. In short, it would be wrong to say that these environments *determine* development levels. For instance, while parts of sub-Saharan Ethiopia are desperately poor, Arizona and Texas in the USA – with a similarly 'extreme' semi-arid environment – are immensely wealthy. Why is this? One reason is that different people respond to similar environments in different ways. People have the capacity to be inventive and, through technology and other means, can to a greater or lesser degree overcome local/regional environmental constraints. This is not to say that people in Ethiopia are not inventive! Rather, it illustrates the point that people have different degrees of *autonomy* in how they respond to local/regional environmental conditions. To understand why development levels vary spatially and temporally, then, we need to consider *both* the environments in question and the specific ways in which different groups use, adapt to and alter them (a point made in Chapter Two by way of the PRT formula): see Figure 7.7 for a graphical representation of the relationships.

3.3 Environmental differences, uneven development and globalization

We are now in a position to link this discussion of environmental differences and uneven development back to globalization. I suggested above that environmental differences help account for, but do not completely explain, variations in levels of human development among places. But this focus on development–environment relationships was primarily concerned with the local or immediate environments that different human groups draw upon. If we want a fuller understanding of the relationships between environment and development we need to turn again to the issue of globalization. This directs our attention to how uneven development, and associated environmental changes, in one place are directly linked to those in another.

As we have seen, environmental differences can help to produce wealth and poverty. Likewise, if these differences themselves change – due to human or natural events – then so too can prevailing patterns of development. I argued above that globalization can alter existing environmental differences or create new ones. It follows, therefore, that *globalization can alter the pattern of uneven development within and between countries by changing the pattern of environmental differences within and between these countries.* Let us return, once more, to our two examples in order to flesh out this claim, using our distinction between 'globalization of' and 'globalization as a cause of' environmental change.

At the start of this chapter we discussed how carbon dioxide emissions from a developed country like the USA can end up being an environmental and social problem for a distant, less developed country like Bangladesh. Although the USA is hardly the only country to emit greenhouse gases, its per capita emission levels – like those of other western countries – vastly outweigh those of developing countries. Put differently, developed countries like the USA are largely responsible for the high levels of greenhouse gases in the global atmosphere (see Figure 7.8). Much of the prosperity these countries have enjoyed since they industrialized has been founded on the mass consumption and burning of resources like coal for heat and light. In effect, one can argue that a country like Bangladesh will pay the price, both environmentally and in human terms, for the development of a set of distant countries. Though a country like the USA has not *intentionally* sought to damage the environments of other nations by burning so much fossil fuel, this damage is the unfortunate outcome of such unrestrained resource consumption. It is a case of what sociologist Anthony Giddens (1984) has called 'unintended consequences'.

If the globalization of environmental change can alter, and even accentuate, the geography of uneven development, so too can the globalization of human relationships. In the case of Amazonian deforestation this might not be immediately obvious. After all, many Brazilians are making a good living from commercial logging. So how can globalization – in this case the connection between Brazil and foreign markets – be said to generate uneven development? The answer to this question leads us to ask once more: what is meant by 'development'? It also leads us to consider the possibility that uneven development can occur *within* a place not just between different places. Let me explain.

In the early 1990s word started to leak out from Brazil that Amazonian logging wasn't benefiting everybody. Though commercial logging companies are the main players in the Amazon's forest drama, they are not the only ones, as I noted earlier. With the assistance of concerned pop stars like Sting (see Figure 7.9), it started to become clear that indigenous tribes in the Amazon were being given very little say in whether and to what extent the Amazon should be logged. These indigenous tribes have inhabited the Amazon for millennia, and certainly before the Portuguese and Spanish forebears of contemporary Brazilians arrived from the sixteenth century onwards. For commercial loggers cutting down trees equates with development: it generates valuable foreign income. For some indigenous Amazonians, however, logging hardwoods is seen as an assault on their spiritual and physical environment. It can hardly be considered 'development' in the positive sense of that term as far as they are concerned. The *same* environmental change – driven by demand in global export markets – has thus *had different implications for different people in the same place.*

You should now have a basic understanding of how globalization, uneven development and environmental change are related. Moreover, it should also be clear why it is important for anyone with an interest in the environment to understand this three-way relationship (Figure 7.7). As we have seen, the

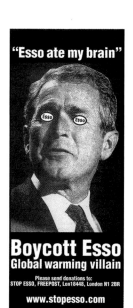

Figure 7.8 In April 2001 newly elected President George W. Bush further reduced the already modest commitments that his country had made to reducing greenhouse gas emissions. He was subsequently accused of having caved in to pressure from big oil firms who stand to lose out financially if the amount of oil purchased and burned is reduced in the interests of mitigating global warming.
Source: The Guardian, 17 June 2001.

Figure 7.9 Elder of an Amazonian indian tribe with Sting, the rock singer, who set up a charity, The Rainforest Foundation.

environmental changes leading to and/or stemming from uneven development in an interconnected world can have serious consequences for the environments people depend upon or value and therefore for those people themselves. To borrow the words of geographers John Allen and Chris Hamnett (1995, p.242), '... uneven development is not just about taking note of difference and diversity, it is about geographical inequality too'. As they also say, '... global relations construct unevenness in their wake *and* operate through the pattern of uneven development laid down' (p.235). In this sense we can talk of **uneven environments**. Different environments and changes to them can have different development implications for different people (and non-humans too) – within a place or between distant places. This is why it is so difficult to make generalizations about the effects of environmental change. As we have seen, it very much depends *who you are* and *where you are* in relation to the environmental changes in question.

uneven environments

Summary

- Environmental differences are a necessary and persistent fact of life on Earth.
- Environmental differences help to cause globalization, in both of the senses we have defined globalization in this chapter, while globalization can alter existing environmental differences.
- People in one part of the world can intentionally or unintentionally be responsible for environmental changes affecting people in other parts of the world, thereby altering existing patterns of uneven development.

4   Globalization, environmental changes and uneven development: so, what's new?

We have now almost provided an answer to two of the three questions this chapter seeks to address. I say almost because we haven't yet fully answered our second question. What's been missing has been a discussion of how globalization, environmental changes and uneven geographical development vary *over time*. Before we turn to our third question, therefore, we need to deepen our temporal perspective on globalization, uneven development and environmental change. You may, at this stage in the chapter, think that the things we have been discussing are relatively recent. After all, the two examples I have been using are both contemporary examples. What is more, the term globalization was coined only a few years ago, suggesting that it refers to processes and events specific to the turn of the new millennium. However, despite appearances, the relationships between globalization, uneven development and environmental changes are far from new. In this section we will briefly explore the history of these relationships. This exploration is worth undertaking, not just because it is inherently interesting. More than this, and as we will see, the *contemporary* causes and consequences of environmental change have a lot to do with *previous* patterns of environmental difference, uneven geographical development and global inter-connection. By implication the *present* situation will directly influence *future* patterns of globalization, environmental change and development.

In the previous two sections you may have noticed that I used the terms 'modern' and 'pre-modern' in my discussion of human–environment relationships. I indicated that anthropogenic global environmental changes are distinctively modern phenomena. But this raises a question: when did the modern era begin? In the context of this chapter, the modern era is not merely synonymous with the last few decades. When do you think it began? One answer you may come up with is that the modern era stretches back some two centuries, to the so-called 'industrial revolution' experienced by western Europe in the early-to-mid nineteenth century. This is, at first sight, a good answer. From this period onwards, countries such as Britain, France and Germany started to do two things. First, they started to alter their domestic environments on a large scale, both intentionally and unintentionally, because they had the technology and population numbers to do so. Secondly, they also started to forge new global connections with peoples in other parts of the world and this had environmental impacts. We will look at some of these industrial-era environmental changes later in this section. But it may surprise you to know that the modern period extends back even further than this. In fact, the globalization of environmental change as a result of human actions goes back some three to four centuries. As you have seen throughout the earlier chapters, humans have been altering their local and regional environments for centuries, but from the seventeenth century these alterations began to become increasingly national and international. To under-stand how and why, let us take a highly illustrative example. Read the case study set out in Box 7.1 and then answer the questions that follow it.

Box 7.1 Environmental change in the North Pacific: the case of fur seals

The year is 1786 and the place is two tiny islands in the middle of the Bering Sea, some 600 kilometres off the coast of what is now Alaska. A few sandy beaches and some scrub vegetation punctuate an otherwise rocky, barren landscape. There's no-one to be seen because the islands are uninhabited by humans. But this does not mean the islands are lifeless. On the contrary, every summer a remarkable natural event occurs. After swimming in the Pacific Ocean for 9 months, some 3 million Northern fur seals (*Callorhinus ursinus*) congregate on just 7 miles of island beach in order to give birth to pups and to mate for the following year. The seals remain on the islands for 3 months before setting off with their offspring for another three seasons in the ocean: see Figure 7.10.

A year later the scene looks rather different. The seals have returned but there's something new about the islands. First, they have been discovered – by Russian explorers – and have been given a name: the Pribilof Islands. Secondly, the two Islands, now called St George and St Paul, each have a small settlement on them. Thirdly, however, the people to be found in these settlements are not just Russians. There are also a few dozen Aleuts, a native people who resemble Eskimos and whom the Russians have moved from the neighbouring Aleutian Islands. Why have these Russians and Aleuts settled on the Pribilof Islands? St Paul and St George are too

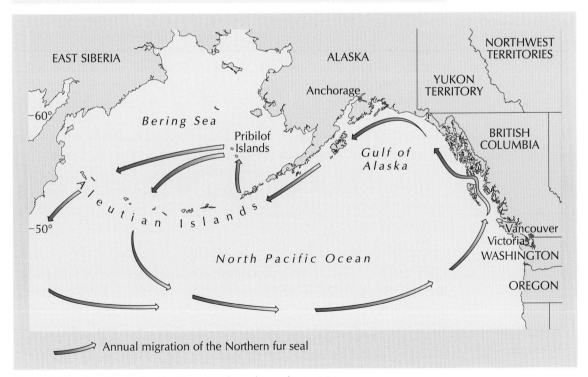

↗ Annual migration of the Northern fur seal

Figure 7.10   The North Pacific Rim: Northern fur seal migration.

small and barren to support agriculture, so food and other provisions must be delivered by ship. So what else would attract people to the Pribilofs? This brings us to a fourth change. Many of the seals that last year occupied the Island beaches are now being rounded up by the Russians and the Aleuts and herded a few hundred metres inland. There they are clubbed over the head and skinned. Why? Because the fur of the seals can be made into luxurious garments thousands of miles away in Russian cities such as Moscow and St Petersburg. In this first year of culling as many as 40,000 seals are skinned. The skins are salted to preserve them and then, when a ship arrives, taken to the other side of the Pacific to be transported by land all the way across Eurasia to Russia. The sealers on the Pribilofs receive a lot of money for their gruesome work. But it is not evenly distributed. The Aleuts are, essentially, enslaved by the Russians. They have been forced to work on the Pribilofs and receive only food, clothing and basic shelter for their labour. They are one of several subjugated groups in the Russian Empire of the eighteenth century.

We now track forward eighty years. Russia is finding it increasingly expensive to maintain an empire and is seeking to pull out of territories it no longer considers to be of political or economic importance. In 1867 it sells Alaska, including the Pribilof Islands, to the USA for $7.2 million. The USA is a young, industrializing country and committed to the principles of free trade. It grants a 20-year licence to seal on the Pribilofs to a newly formed private firm, the Alaska Commercial Company (the ACC). This company replaces most of the Russian sealers with its own men, but retains the Aleuts because they are skilled at clubbing, skinning and salting (Figure 7.11a). By this time the Aleuts number 600 and the settlements on St Paul and St George have schools, churches, law and the like. Over its 20-year lease the ACC kills 90,000 seals per year for a profit of $27.8 million (almost 4 times the 1867 cost of the whole of Alaska!). From 1890 another company, the North American Commercial Company (the NACC), is awarded a licence to seal by the US government. It continues the practices of the ACC, but is able to kill just 17,000 seals per year for a profit of $4.4 million during its two decades of operation. Over the 40 years of American sealing the Aleuts, who are now paid a wage and are no longer enslaved, earn some $1 million in total salary. The 100 Americans on the islands, by contrast, earn five times more per person than the Aleuts.

The Pribilof furs are no longer destined for Russia but for New York, London and Paris, where wealthy urban consumers wear them as hats and coats. But, as noted, the number of seals killed declines steadily as the nineteenth century gives way to the twentieth. Why is this? The main reason is competition. In 1857 Canadians arrive on the coastline south of Alaska and create the province of British Columbia. One of the things they immediately notice is the large number of fur seals that swim north along the coast between January and May every year. By 1890 there are around 100 schooners, or sealing ships, catching seals at sea and operating from

Figure 7.11   (a) Killing fur seals on the Pribilof Islands in the late nineteenth century.

Figure 7.11   (b) Fur seal carcasses, St Paul, *c.*1939.

the port of Victoria. The Americans are unable to stop this ocean sealing because it occurs outside the territorial waters of Alaska (refer back to Chapter Five for more on the law of the sea). Moreover, because this sealing at sea is unregulated by the Canadian government, it is unclear how many seals are now being killed each year. What is certain, though, is that fewer and fewer seals are making it to the Pribilof Islands to breed. In 1911, therefore, the USA and Canada sign an agreement to control sealing before the seals become extinct. Part of this agreement involves banning all sea sealing in perpetuity, leaving land sealing on the Pribilofs the only permissible way to kill Northern fur seals (Figure 7.11b).

We now track forward once again, to the present day. By 1911 there were just a few tens of thousands of seals making it to the Pribilofs. Today there are almost 3 million, as there were prior to 1786. The Islands are still part of the USA but commercial land sealing was banned by the US government in 1967. Part of the reason for this ban was that social attitudes towards killing wild animals changed. In the US and other countries, growing environmental awareness from the early 1960s meant that governments came under pressure to outlaw the mass killing of mammals such as seals. That awareness persists to this day. On St George there is now a Seal Conservation Centre run by the US Wildlife Service. The Centre studies the behaviour of Northern fur seals and is dedicated to their conservation. The Centre is visited by around 20,000 tourists per year, most of them from Canada, mainland USA and Japan. These tourists visit the Pribilofs during the summer to see seals mating and birthing. This tourist trade sustains the 350 people who today live on the Pribilofs. Many of these Pribilovians are direct descendants of the first Aleuts, and they own hotels, tour boats and restaurants. Their standard of living is similar to that of the average American. Since 1972, these 'native Americans', as they are now called, have joined with other native groups in Alaska in an attempt to win control of the Islands for themselves. Their argument is that because they lived on the Islands before the Americans arrived in 1867 they should be entitled to ownership of the land and all monies relating to seal tourism.

Source: Much of this story is adapted from Busch-Cooper (1985).

Activity 7.6

The story of Northern fur sealing that you have just read is as interesting as it is illustrative. By answering the following questions we can begin to see what a fascinating case it is, one that tells us some important things about globalization, uneven development and environmental change in the past and their relevance to the present:

1   Of the two types of global environmental change identified in Section 2.1, which category does the decline and rise of the Northern fur seal population fall into?

2   Over time, how has globalization caused changes in the numbers of seals arriving in the Pribilofs?

3   Historically, how has the changing number of fur seals affected the levels of development on the Islands?

Comment

The answers to these questions, except perhaps the first, are neither short nor simple. Let us take each in turn.

1   The changing fortunes of Northern fur seals are an example of a local environmental problem with global causes (though, in this case, not necessarily global consequences, unless one considers the near-extinction of the seals in the 1910s a potential loss to global humanity or their current conservation a global gain).

2   What's particularly interesting here is that while the globalization of human activities is key to explaining this local environmental change, the *form* of globalization has altered over time. In the 1780s it was environmental difference – that is, the fact that so many seals congregated on these islands and nowhere else – that caused the Russians to settle the Pribilof Islands. But this early form of globalization, linking two tiny knuckles of rock in the Pacific with consumers half a world away, was distinctively *imperial*. **Imperialism** was a process whereby one nation would conquer and occupy another, extracting resources from it. Russia was one of several European nations that, from the fifteenth century, founded empires and increased their wealth on the basis of this conquest. The other nations included France, Germany, Great Britain, Holland, Portugal and Spain, who between them at one time had empires covering half the non-western world.

Russian occupation of the Pribilofs led to the sudden annual decline in the number of seals, all because of the demand for their fur among wealthy Russians. However, because Russia had a monopoly on the Pribilof seals – meaning that non-Russians either did not or could not kill them – they could at least control the number of kills. This changed, as we saw, when the Americans took over in 1867. American involvement in the Pribilofs not only changed the geography of the Pribilof's global connections, which were now to fashion designers and consumers in the eastern USA and western Europe. More than this, it involved a related but slightly different form of globalization, a *capitalist* form based on economic competition and profit. Imperialism was linked to **capitalism** insofar as Russia (in this case) used its political control of the Pribilofs to take seal skins to distant markets where they were sold for profit. But capitalism can operate without imperialism, as became clear once the Russians left Alaska in 1867. As you saw from reading the case study, neither the ACC nor the NACC could stop Canadians from killing seals at sea before they reached the Islands. Both the companies and the sea-sealers were competing over the same seals in order to serve the same foreign markets. The rational way for both sides to maximize the money they could make from sealing was to kill as many seals as they could before the other side got them. Inevitably, this scramble for seals between groups in different places (the Pribilofs and Victoria) led to their precipitous decline during the 1890s and 1910s. Since the 1960s, by contrast, the spread of 'post-industrial' 'environmental values' in North America and Western Europe has altered the image of seals from a 'resource' to be used for human benefit to a valuable life-form which deserves the right to live free from harm (this relates to the instrumental and non-instrumental values discussed by **Hinchliffe and Belshaw, 2003**).

Imperialism

capitalism

3 Since 1786 the rise and fall of seal numbers has been directly linked to levels of development on the Islands. As seal numbers declined and rose, so too did the incomes of those on the Pribilofs. By the 1910s, meeting the fashion needs of consumers in London, New York and Paris was starting to endanger not only seals but the livelihoods of Pribilof sealers. Today, by contrast, the development of the Pribilofs is sustained primarily not by distant strangers but by tourists coming physically to the Islands to spend money on seal-watching and related environmental activities.

However, there is a layer of complexity that needs to be considered here, reminding us again that uneven development occurs *within* as much as *between* places. Historically, the Pribilof Aleuts have not enjoyed the full benefits of sealing. Post-1786 occupation of the Islands by Russians entailed the forcible relocation and control of the Aleuts. Prior to Russian conquest the Aleuts had been a nomadic people who roamed the Aleutian Islands in search of fish, seals and other marine life. After 1786 the Pribilof Aleuts had to adjust to being confined to two foreign islands, to working for Russian overlords and to slaughtering *en masse* seals they had hitherto killed in small numbers only for food. Even when the Americans arrived in 1867, and imperial rule formally ended, the Aleuts fared only marginally better, receiving a disproportionately low share of total profits earned from sealing. It is only in the last few decades that this native people have received a fairer share of the income deriving from sealing. But as their current attempt to win political control of the Islands shows, development cannot just be measured in monetary terms. For the Aleuts it is also about the right to control their own land free from 'foreign' (in this case US) influence. Because of their group values they see sealing and its legacy in different ways to other Americans.

To conclude, what this analysis of sealing on the Pribilof Islands shows is the *varying* links between environmental change, uneven development and globalization over time. The *same* place can find itself experiencing *different* degrees or types of environmental changes over time because of *different* forms of global human relationships resulting in uneven development within and beyond the place. What is more, *previous* rounds of environmental change directly influence *present* patterns of environmental change. Though the case of the Pribilof seals is only one of many examples we could have considered, we can certainly learn wider lessons from it.

Summary

- The links between environmental change, uneven development and globalization are not new.
- In order to understand environmental changes, uneven development and globalization today we need to understand history and the continuing effects on present forms of activity.

5   Explaining anthropogenic environmental changes

We have seen in this chapter why comprehending the human causes of environmental changes matters. If we only look at the physical causes, we get an incomplete picture of what is going on. We have also seen why understanding the human consequences of these changes matters. Environmental changes do not have to be life-threatening in their outcomes in order to be taken seriously. The case of Amazonian natives or Aleutian sealers shows that the economic or spiritual losses (or gains) resulting from environmental changes can be very important for certain groups of people. While a given environmental change might be beneficial for some, it might represent a problem or a loss for others. As I intimated in the introduction to this chapter, environmental changes can produce winners and losers (as captured by the idea of 'uneven environments').

We are now in a good position to answer the third main question that this chapter seeks to address. Just to remind you, the question is this: how can we model (i.e. represent and explain) human-caused environmental changes and their consequences in a world of globalization and uneven development? Thus far we have been rather descriptive in our argument. So now we need to specify what the *root causes* are of some of the things we've been describing. It should, you may think, now be a fairly straightforward task to identify some of these root causes. However, we here encounter an interesting problem. In the discussions of global warming, Amazonian deforestation and Pacific sealing, I have implied that the human causes and consequences of environmental change are clear-cut and can be agreed upon by those studying them. But this is, in fact, a problematical thing to imply. For as I indicated in my introductory comments, there exists *disagreement* about what the human causes and consequences of environmental changes actually are. So far, I have kept these disagreements hidden from view, but now we need to consider them head-on. We need to do so not just for academic reasons but for practical ones as well. Only by having a clear understanding of the human causes and effects of environmental changes can we address any problems that might arise from these changes. In this endeavour social science, together with natural science, can help to devise practical policies to alter human actions (**Blowers and Hinchliffe, 2003**). But if analysts differ in their understandings, then things get slightly complicated. Two different analysts examining the *same* environmental changes may come up with *different* accounts of its human origins and outcomes. In turn, this might mean that each analyst will suggest quite different policy solutions to any problems arising from this environmental change – with one analyst even perceiving a problem where the other sees no problem at all!

To put some flesh on these rather general arguments, in this section we will look at two important approaches to understanding the human dimensions of environmental change: the **tragedy of the commons** and **livelihoods** approaches. In brief, the livelihoods approach looks at how the *economic context*

tragedy of the commons
livelihoods

in which people make a living influences their relationship with local and non-local environments, while the tragedy of the commons approach looks, more specifically, at people's *property rights* over environments (the latter was touched upon in Chapter Four). Before we compare them in detail, though, we first need to look briefly at what the two approaches have in common.

5.1 Models of environmental change

Although I have used the general term 'approaches' to describe them, the tragedy of the commons and livelihoods perspectives are better described as models. You have already encountered this term in earlier chapters, but it's worth reminding ourselves what models are and why they are useful when studying the world. Both social and physical scientists devise and use models when they study the world. The word 'model' might immediately conjure up images of a fashion 'super-model' or a plastic aeroplane kit! But in both the physical and the social sciences the word model means something else. A model, in this specialist sense, is a simplified representation of reality that scientists construct in order to better understand how that reality works. Specifically, models do two things:

- they allow scientists to 'filter out' detail and complexity in order to identify the basic workings of a situation;
- they can be applied to many apparently different situations.

So models aim for both clarity of explanation and a certain universality of application. In principle, it is possible to develop a model to represent and explain any real-world situation occurring at pretty much any scale. For scientists models are both useful and efficient. They are useful because they allow scientists to see or at least infer the fundamental processes at work by eliminating incidental detail. And they are efficient because once developed they can be used to study many cases of the thing they refer to. Physical scientists have produced many, now famous, models to explain the natural world. Examples include biologist Charles Darwin's model of the evolution of life-forms and physicist Albert Einstein's model of the relativity of time and space. In social science, the Malthus model is widely known (and you were introduced to it in Chapter Three) but the tragedy of the commons and livelihoods models slightly less so. It is to these two models that we now turn. As we will see, what they share is a preoccupation with the *economic dimensions* of human life and human decision-making. However, they differ in how they depict the nature and spatial scale of these dimensions.

5.2 The tragedy of the commons model

The tragedy of the commons model is associated with the US scientist Garrett Hardin (1968) and is a 'highly influential explanation of a wide range of humanly produced environmental problems' (Barrow, 1999, p.231). Hardin was concerned to identify the underlying or fundamental reasons for a range of otherwise different environmental changes. He argued that these reasons related to the ownership – or lack of ownership – of environments and to the logic of economic decision-making in the modern world. People use and value environments in multiple ways – practical, spiritual, moral and aesthetic, for example – but at some level most environments are used by humans for economic purposes. By **'economic'** I mean the production of goods and services for personal or communal consumption and/or sale to others. With this in mind we can understand the elegant logic of Hardin's model.

economic

Hardin argued that in the modern world what is economically rational for individuals and communities often turns out to be environmentally irrational. To illustrate this contradiction, Hardin took the case of a fictional group of farmers herding their cattle on a common pasture. The pasture-land is free for all herders to use, though is finite in its carrying capacity. Let us suppose there are 20 farmers, that each owns 1 cow, and that the pasture can support the herd comfortably. For Hardin this state of affairs does not last long because it is economically rational for each farmer to try to add as many more cows as possible to the pasture. Take the case of farmer A (see Figure 7.12). She might buy an extra cow at market. Why? Because she would have twice the milk and meat whilst the extra grazing pressure of adding this cow will be borne not by farmer A alone but by *all* other farmers. Thus farmer A enjoys 100 per cent of the benefits of the extra cow but only 5 per cent of the disbenefits! However, the problem is that if *every* farmer does this in order to gain *individually*, then the eventual outcome will be over-grazing leading to gradual herd and economic decline *for all*.

The root of this 'tragic' scenario for Hardin lies in the lack of property rights in the environment. In the scenario above, while the farmers own their cows they do not own the pasture-land those cows graze on. It is a 'common' or 'open access' resource, owned by no one person but free to be used by everyone. For Hardin

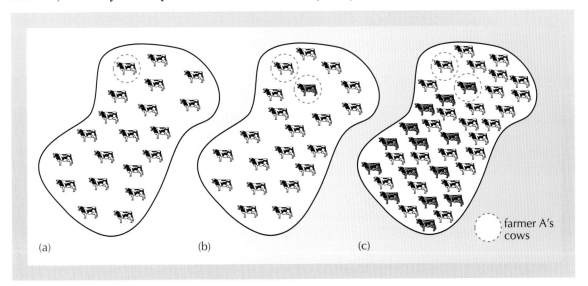

(a)   (b)   (c)

farmer A's cows

Figure 7.12 The tragedy of the commons.
(a) Each farmer begins with one cow; (b) however, if we take the perspective of an individual farmer – say farmer A – it is rational to immediately add an extra cow; (c) if only farmer A adds an extra cow there is no problem, but in practice it is rational for every farmer to add a cow, thus doubling the herd and leading to overgrazing and economic ruin for all.

Hardin's model applies to the oceanic over-fishing discussed in Chapter Five of this book.

this 'missing property right' means that no-one takes responsibility for the proper management of the pasture. Instead, there's a free-for-all in which each farmer, acting rationally as an individual, tries to derive the most economic benefit from use of the pasture. Of course, this free-for-all only occurs if the farmers do not communally regulate their use of the pasture.

Activity 7.7

Having digested the tragedy of the commons model, see if you can now apply it to the three examples of environmental change that we've been focusing on thus far – global warming, Amazon deforestation and seal harvesting. For each case, can you identify how what seems an economically rational course of action ultimately turns out to be environmentally irrational? In each case you will need to identify which part of the environment is open-access and thus free to be used in an unregulated way.

Comment

Did you find the activity difficult or straightforward? Though seemingly different in their nature and causes, our three examples can all be seen as 'tragedies' in Hardin's sense of the term. In brief:

- In the global warming case, people worldwide release greenhouse gases into the atmosphere because they are free to do so. Since the atmosphere is not owned by anyone people can pollute with impunity. Indeed, it is economically rational to pollute! For instance, a power company burning coal will emit CO_2 into the atmosphere because there is normally no charge for doing so. Since no-one owns the atmosphere, there is no-one to compensate for the pollution caused. Yet in the long term, global warming will affect everyone on the planet, polluters included, albeit in different ways and with different consequences across space and over time.
- In the case of the Amazon things are slightly more complicated in that logging companies either own or have leased the land they log. So the *forest* is not an open-access resource like the pasture in Hardin's fable. How, then, does Hardin's model apply to this case? The tragedy here concerns the relationship between the forest and the atmosphere. While it may be economically rational – because profitable – for logging companies to deforest the Amazon, these companies are not penalized financially for removing one of the world's major natural sinks for CO_2. Again, the problem here is that the atmosphere is not owned. In the absence of government regulation, logging companies pay little heed to the long-term atmospheric effects of removing forest because these effects are free to be ignored economically.
- In the case of seals, Hardin's model applies in fairly clear-cut ways. Here it is the seals that are an open-access resource since neither the Americans nor Canadians owned them. Knowing that the seals would end up on the Pribilof Islands each summer, the Canadian sea-sealers of the late nineteenth century tried to catch as many seals as possible before they became American 'property' by entering Alaskan waters.

So, the tragedy of the commons model seems to offer a convincing and widely applicable explanation of why people cause environmental changes that may ultimately create problems for themselves or for others separated across space and through time. It implies one or all of three solutions to avert 'tragedy'. This is the normative side of the model. From our examples, can you see what the solutions are? One is to privatize environments so that the new owner takes responsibility; a second is for different users to develop a communal management system; and a third is for governments to step in to control how environments are used by economic actors. Paul **Anand (2003)** discusses these solutions in more detail. What they have in common is that they focus on altering property rights over environments and/or the degree (but not necessarily the *way*) to which these environments are exploited. However, the tragedy of the commons model is not the only one that has been proposed to explain anthropogenic environmental change in a world of interdependence and uneven development.

5.3 The livelihoods model

The livelihoods model is less concerned with people's property rights over environments and more interested in how they make a living. Examining livelihoods can tell us an awful a lot about the nature and impact of people's relationships to their environments (and my discussion of livelihoods in this and the next section overlaps with the $I = P \times R \times T$ equation discussed in Chapter Two – see if you can identify how). The livelihoods model is a synthesis of old and new ideas developed by radical political economists like Karl Marx (in the nineteenth century) and Amartya Sen (presently of Cambridge University) (see Figures 7.13 and 7.14). But what are 'livelihoods'? The term refers to something much broader than the jobs people do: it refers to the whole complex of means whereby persons obtain the necessities of life, and can include considerations (Bernstein, 1992), such as:

Figure 7.13
Karl Marx (1818–1884) remains among the most influential political economists of modern times. His critical analysis of capitalism, unfolded in the several volumes of his book *Capital,* have influenced several generations of scholars and policy-makers. Marx's fundamental belief was that capitalism generates social and geographical inequality and that a socialist economic system was a more humane and environmentally friendly alternative.

- who owns what or has access to what?
- who does what?
- who gets what?
- what do they do with it?

As we will see in the next and penultimate sections, we can analyse livelihoods at two levels and in two ways.

First, because many different people make their living worldwide in the same ways, we can inquire into the nature of the economic systems they commonly find themselves in. By 'economic system' I mean the rules that govern how and why goods and services are made and delivered to others. Not all economic systems are the same and the question of which system to embrace has, historically, been a question of politics and power for different nations and places. Since people interact with environments in ways that are structured by these systems, the livelihoods model can tell us a lot about the nature and consequences of this interaction.

Figure 7.14
Amartya Sen is a Nobel Prize winning economist and a Professor at the University of Cambridge. His investigations of how famines can occur in the midst of food surpluses led to his concept of 'entitlements'. This important idea focuses on how individuals and families often lack the monetary and non-monetary resources to acquire food. Importantly, entitlements, in Sen's view, are partly determined by social relationships that are beyond the control of the individuals and families who possess them.

Second, economies are political and entail power relations because they involve questions of who should enjoy the benefits of economic growth or suffer the costs. As we have already seen in this chapter, inequalities between people – in the same place or between places – can be a direct result of their relationships with one another. Here, then, the livelihoods model looks not just at the general rules defining economic systems but at how, and with what effects, incomes are distributed within them between different groups. In sum, in this model 'it is livelihoods that establish the relationship of human populations with their environment' (Woodhouse, 2000, p.149).

We will explore this model in more detail in the next section. For now, though, it's worth briefly comparing it with the tragedy of the commons model. Both models are concerned with how economic activity affects environments. As we will see, the issue of property rights is not irrelevant to the livelihoods model. Rather, this model argues that property rights are not the *fundamental* reason why people alter environments in the ways they do. In this model, while property rights do indeed help us to understand why problems such as global warming and over-sealing occur, they fail to consider the wider economic systems of which property rights are just one element. This is why, in the remainder of the chapter, I want to complement the tragedy of the commons model with insights provided by the livelihoods approach.

Summary

- Identifying the human causes of environmental change is complicated by the fact that there is a range of models one can draw upon.

- Social scientists, like their physical science counterparts, use models to explain environmental changes.

- The tragedy of the commons model argues convincingly that property rights over environments are a key element in explaining a range of anthropogenic environmental changes in an interdependent and uneven world.

- The livelihoods model is less concerned with property rights over environments than with the nature of, and group relationships within, economic systems.

6   Livelihoods and environments in a global economy

Livelihoods were defined above as the particular way in which people make a living. All livelihoods have environmental effects. Directly or indirectly they involve (a) using raw materials (such as Amazonian trees) and/or (b) using the environment to accept waste products (as in the case of atmospheric pollution). As we have seen with our consumers of luxury fur seal garments and mahogany furniture, even someone whose livelihood seems far removed from the natural environment can have a serious impact upon it. This raises the question of how best to categorize livelihoods so that we can judge their role in environmental change. One obvious thing to do is to look for livelihood types (is a person a miner, a stockbroker, a logger etc.?). However, livelihoods analysts see this as 'missing the wood for the trees'. They ask instead: what do these different types of livelihood have in common? The answer, they argue, lies in the nature of the economic system or systems that these different livelihoods are together embroiled in.

6.1 Economic systems and environmental changes

What is an **economic system**? It is a particular way in which a society meet its economic requirements. All economic systems have two dimensions. First, they have an overall aim. Secondly, they involve specific kinds of relationships between people. Together these two things comprise the 'rules of the game' for how otherwise different people make a living. Let us take each in turn in this and the following sub-sections.

economic system

If we think again about Pribilof sealing, we saw two economic systems at work simultaneously: an imperialist system and a capitalist system. Prior to 1786 the Aleuts had their own subsistence system, which we did not consider. Can you recall the key differences between the imperialist and capitalist systems? The imperialist system was based on extracting resources from dominated lands and peoples in a highly exploitative way. Its aim was to transfer goods to countries such as Russia while giving little or nothing back to the exporting areas. The development of the ruling country was very much bound up with the lack of development of the ruled. The capitalist system, by contrast, was based on competition between different economic actors with equal rights to undertake the economic activities they did. Its aim was to gain the maximum profit for those agents in a predominantly 'free market' situation where consumers could choose to buy or not buy commodities (like seal furs). In fact, I am wrong to use the past tense here: for capitalism is still very much with us today. Even those environmentally benign tourists, who today pay money to watch seals in the Pribilofs, are buying a product – 'environmental tourism' – from economic actors seeking to make a profit (hoteliers, restaurateurs and boat-owners on the Islands). So capitalism is about selling things, whatever they may be, with the principal aim of making money. This is what Karl Marx called 'accumulation for accumulation's sake'. Capitalism's overriding 'logic' is economic growth.

A good way to appreciate the distinctiveness of capitalism is to reflect on the subsistence economic system that many pre-modern peoples developed (like the Aleuts). In this system goods were produced not for profit but simply to meet basic, everyday requirements for food, shelter and clothing. Moreover, subsistence systems were generally local in scale. By contrast, capitalism, and its imperialist predecessor, were/are global systems. Capitalism, as we have seen through the examples used in this chapter, links producers and consumers in far-flung parts of the world. In fact, capitalism is arguably *the* dominant economic system in the world today. People all over the planet doing all manner of different jobs have one thing in common: they are involved in a system where things are sold primarily in order to make money. The global links between these people are not incidental to capitalism. Rather, as we have seen with seals and Amazonian trees, these links keep the system going. Without distant markets many local producers could not make a living.

So the livelihoods model encourages us to look first at the specific *aims* of an economic system. We can, in turn, relate this overall aim to environmental changes, finding common patterns in apparently diverse situations worldwide. If we take capitalism both yesterday and today we can see how environmentally *destructive* it can be. More precisely, the rules governing how goods and services are produced within capitalism often encourage people to degrade environments directly or indirectly through their economic actions. This destructiveness operates in two main ways. First, the case of sealing showed how *competition* between the ACC, the NACC and Canadian sealers led to a near-catastrophic decline in seal numbers – principally in order to reap profits from selling luxury garments to acquisitive, wealthy consumers thousands of miles away. This economic competitiveness is a key way in which the capitalist economic system functions. Secondly, and somewhat differently, global warming and Amazonian deforestation show us another side of capitalism, one less about the effects of competition among economic actors and more about *how the quest for profit can override non-profit considerations.* From a livelihoods perspective, one of the main reasons why businesses worldwide pollute the atmosphere, or fail to pay heed to how forests like the Amazon serve as important sinks, is because protecting the environment is not the principal aim of capitalist enterprise. If businesses were required to pay for their pollution, or to prevent it by using appropriate technology (and they sometimes are forced to do this by law), the additional costs would eat into their profits. So, within a strictly capitalist system, it is 'rational' to pollute, despite the fact that the environmental effects may be harmful. It is not that people making a living in a capitalist system *want* to pollute the environment. It is more that the 'rules' of this system make it logical to save money/make greater profits by polluting. In the tragedy of the commons model this is attributed to the presence or absence of property rights over environments. But in the livelihoods model, property rights are the *symptom* where the *underlying cause* lies in the rationale for producing goods and services.

In sum, in terms of its overall aims, capitalism can be an environmentally costly way for people across the planet to make a living. These aims have led to all manner of harmful global environmental changes with local causes and consequences and local environment changes with global causes and consequences. These changes are, in effect, the environmental costs of the continued economic growth that is the fundamental aim of the capitalist system. Left unchecked, the capitalist principle of 'accumulation for accumulation's sake' means that environments are used to the point of destruction to satisfy the thirst for profit. In a capitalist world, therefore, the general *form* of human–environment relationships is geared to humans using the environment as *a vehicle for making money*. This said, we should not think that the aims of capitalism are necessarily destructive of the environment. Sometimes it can be profitable to protect environments – as with ecotourism, for example – and in other situations governments can place restrictions on the environmental damage caused by economic activity (see **Anand, 2003**). Moreover, non-capitalist systems can also be highly destructive of environments – a good example being the former communist economies of the USSR and eastern Europe.

6.2 Environmental changes and human inequality in a capitalist world

So much for the first aspect of an economic system. What about the second? This involves looking at the economic relationships between people. These relationships are a distinguishing feature of economic systems. Consider imperialism again. We saw how the Russians forced the Aleuts to work in the seal trade, benefiting from their free labour while reaping the full economic rewards of selling seal pelts. In capitalism, by contrast, production is organized as follows. A businessperson pays workers to make or gather commodities, and these commodities are sold to consumers in local, national or global markets for a

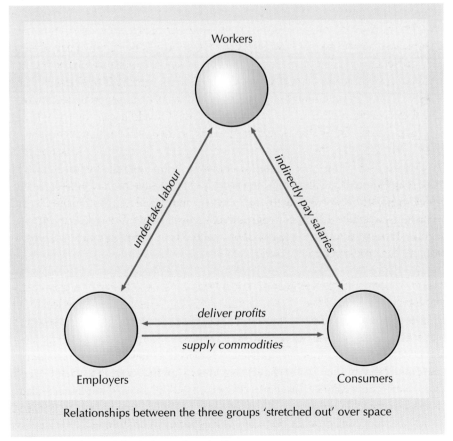

Relationships between the three groups 'stretched out' over space

Figure 7.15 The social relations of capitalism.

profit. Clearly, part of a businessperson's profits are used as wages. So capitalism does not involve formal coercion in the way that imperialism did. This said, the social relationships that sustain this system do have some potentially contra-dictory characteristics. Firstly, the objectives of the three principal groups in the system – owners, workers and consumers – are not necessarily harmonious. For instance, owners might want to maximize profits, workers will want higher wages, and consumers will want lower prices – a contradictory triangle since it is unlikely that all these objectives can be achieved to each group's satisfaction simultaneously (Figure 7.15). Secondly, because capitalism is a global system, the objectives of any individuals in one of these groups will be influenced by where they are geographically. As part of this 'where' issue we need also to consider how the past influences the present, the point I made in Section 4 of this chapter. So livelihoods vary depending on *who* you are and *where* you are. Once we add these two distinguishing features of human relationships within capitalism together, we can understand the second dimension of how the livelihoods model explains environmental change and its consequences in the modern world. Let me elaborate on this by way of a final example.

Activity 7.8

Read the case study in Box 7.2 and adopt a livelihoods perspective. See if you can identify two things. First, can you see how the long-distance connections between different economic actors have caused environmental changes? Second, can you see how the impacts of these changes fell unevenly on these actors depending on who they were, where they were and what their economic interests were?

Box 7.2 Groundnut cultivation and environmental change in Niger

In the 1950s, France, like Britain, still had colonial territories in the 'Third World'. Like other European countries it consumed a lot of vegetable oils and was heavily reliant on imports of oil sources, such as soya beans, from the USA. Alarmed at its dependency on imports from one country, France sought its own sources of vegetable oils.

In Niger, one of its African colonies, France expanded groundnut production 100 per cent between 1954 and 1957 alone. When Niger gained independence from France in 1961 it continued to export groundnut oil and by 1968 had expanded production a further 50 per cent. In order to find the extra 130,000 hectares of land necessary to meet French demands for oil, the Niger government encouraged groundnut production on soils that had hitherto been left fallow and in areas of low rainfall. These low rainfall areas had been used for generations by roving pastoralists, who now came into conflict with groundnut producers as their pasture land was brought into cultivation. Between 1968 and 1974 several droughts put these pastoralists under enormous pressure and they lost around 40 per cent of their livestock. Meanwhile, groundnut production proceeded relatively smoothly – courtesy of irrigation measures.

Figure 7.16 (left) Groundnut farming in Niger, and (right) pastoralists in Niger, herding sheep.
Source: Source: Woodhouse, 2000.

Comment

What are your answers to the two questions? The switch to groundnut production was not only a notable environmental change in its own right; it also altered the environmental conditions of pastoralists. Driving this specific change in environmental use in Niger – a poor African country – were changes in market demand in France, themselves caused by national government reaction to US domination of vegetable oil exports. Here we have a local or regional environment change with global human causes. But the changes arose from a *combination* of two types of globalization: imperial (France had the power to force through groundnut production in Niger in the 1950s) and capitalist (France reacted to the US's market dominance in vegetable oils by developing its own supplies). After Niger's independence it became a market competitor with the US in its own right.

So the 'distanciated' connections between vegetable oil consumers and producers, mediated by imperial power and capitalist competition, were a decisive cause of the expansion of groundnut production in Niger. What, though, of the group impacts of this environmental change (our second question)? Clearly, the *nature and significance* of these changes varies depending on who you are, where you are situated and what your economic interests are. Consumers of vegetable oil benefited not just from its availability but its affordable price. Groundnut oil was generally cheaper than oil derived from US soya in the 1950s and 1960s. Likewise, groundnut producers in Niger benefited from oil demand in France, while their workers, though far less well-off financially, managed to earn a basic living working on groundnut plantations. However, pastoralists in Niger were, clearly, made much more vulnerable by the increase in acreage devoted to nut production. **Vulnerability** is a measure of how   vulnerability susceptible a person or a group is to the negative impacts of environmental changes (see Chapter Six). Pushed into more marginal environments, the pastoralists' capacity to withstand drought was reduced, with ultimately dire consequences for themselves and their cattle. Only those few pastoralists with extra assets – such as money or a means of living other than through cattle alone – survived drought relatively unscathed (they were more 'resilient' than less well-off pastoralists). So, from a livelihoods perspective, the *same* environmental change (in this case a turn to groundnut production) can have *unequal* human consequences.

This last observation leads us to a final important insight deriving from the livelihoods model. If certain groups or places are unable to sustain their livelihoods then they may over-exploit environments, which in turn may exacerbate or merely postpone their vulnerability. In the case of the Niger pastoralists, they were forced to graze dry, sparse pastures in order to maintain their livelihood which, under drought conditions, made the pasture even more marginal over time. This spiral of environmental decline ultimately worsened the pastoralists' economic situation and threatened their way of life.

What can we learn from this? Two things, it seems to me. First, when people's livelihoods are threatened, environmental changes are often set in train that exacerbate their poverty. So poverty can be at once an outcome and a cause of environmental changes, implying that environmentally 'sustainable' development must eliminate poverty (see **Hinchliffe and Belshaw, 2003**). Second, in an interconnected world, we need to ask who should take responsibility for preventing harmful forms of environmental change. In the Niger case, we saw that at some level the pastoralists' plight was due to the actions of others – groundnut farmers, governments and distant consumers. This, from a livelihoods perspective, raises complex issues of who should assist vulnerable groups such as Niger pastoralists when their immediate environments are degraded. This is a more technically and morally complex issue than simply altering property arrangements or the rate of environmental degradation as the tragedy of the commons model suggests.

Summary

- Economic systems vary in nature geographically and historically and have a range of environmental effects.
- The capitalist economic system can produce both categories of global environmental change identified earlier in this chapter.
- In a capitalist economic system, environments become vehicles for making money, while different people in this system enjoy or suffer environmental changes depending on who they are, where they are and what their livelihoods are.

7   Conclusion

In this chapter we have answered three major questions about globalization, uneven development and environmental changes. You should now understand the ways in which local or regional human actions can have environmental impacts at larger and longer spatial and temporal scales. You should, secondly, be able to explain why uneven development across space and over time is both a cause and an outcome of these large-scale environmental changes. Finally, you should be able to compare and contrast two of the principal models available to determine what it is about human activities in the modern world that produce these environmental changes. Together, these three learning outcomes will hopefully demonstrate to you just how *complex* are the causes and effects of different environmental changes at different scales in a world where people's needs and wants are so many and varied. In this world of uneven development and global relationships between people, we've learnt that environmental changes can have serious and inequitable consequences. This inequity helps to explain why so many people today actively struggle to improve local and non-local environments, for both themselves and others. This struggle and contestation is the subject of the next book in this series (**Bingham et al., 2003**).

References

Allen, J. (1995) 'Global worlds' in Allen, J. and Massey, D. (eds) *Geographical Worlds*, Oxford, Oxford University Press/The Open University.

Allen, J. and Hamnett, C. (1995) 'Uneven worlds' in Allen, J. and Hamnett, C. (eds) *A Shrinking World?*, Oxford, Oxford University Press/The Open University.

Anand, P. (2003) 'Economic analysis and environmental responses' in Blowers, A.T. and Hinchliffe, S.J. (eds).

Barrow, C. (1999) *Environmental Management*, London, Routledge.

Bernstein, H. (1992) 'Poverty and the poor' in Bernstein, H., Crow, B. and Johnson, H. (eds) *Rural Livelihoods: Crises and Responses*, Oxford, Oxford University Press/The Open University.

Bingham, N., Blowers, A.T. and Belshaw, C.D. (eds) (2003) *Contested Environments*, Chichester, John Wiley & Sons/The Open University (Book 3 in this series).

Blowers, A.T. and Hinchliffe, S.J. (eds) (2003) *Environmental Responses*, Chichester, John Wiley & Sons/The Open University (Book 4 in this series).

Busch-Cooper, B. (1985) *The War Against the Seals*, Kingston and Montreal, Queens–McGill University Press.

Giddens, A. (1984) *The Constitution of Society*, Cambridge, Cambridge University Press.

Greenpeace (2001) *Save or Delete? A Last Chance to Save the World's Ancient Forests*, Amsterdam, Greenpeace.

Hardin, G. (1968) 'The tragedy of the commons', *Science*, vol.162, pp.1243–8.

Hinchliffe, S.J. and Belshaw, C.D. (2003) 'Who cares? Values, power and action in environmental contests' in Hinchliffe, S.J., Blowers, A.T. and Freeland, J.R. (eds) *Understanding Environmental Issues*, **Chichester, John Wiley & Sons/The Open University (Book 1 in this series).**

IPCC (Intergovernmental Panel on Climate Change) (2001) *Third Assessment Report*, Cambridge, Cambridge University Press.

Woodhouse, P. (2000) 'Environmental degradation and sustainability' in Allen, T. and Thomas, A. (eds) *Poverty and Development in the Twenty-first Century*, Oxford, Oxford University Press.

General conclusions: thinking about environmental change

Dick Morris

Now that you have read through the chapters of this book, you should have a clearer idea of the ways in which environments change, and the processes that cause these changes, providing some answers to the first key question posed in the introduction: how and why do environments change? You should also have become aware of the importance of time and space in considering environmental change. Change occurs over many different scales of time, from the almost instantaneous to those that, to a human, are almost unimaginably long. You should now recognize that we often need to look at processes over this whole range of timescales in order to understand what is happening to environments at the present moment. So, we may need to examine the instantaneous interception of light by a leaf, the rapid chemical reactions of photosynthesis, through the longer-term successional changes in vegetation communities, the even slower evolutionary changes both in species and in human technology and culture, right through to the geological processes of plate tectonics and even planetary formation (Chapters One and Two). We also need to understand processes at a range of spatial scales – from the individual leaf, the biotic community and landscape, intercontinental migratory movements of living organisms, the ocean currents to the whole atmosphere (Chapters Five and Six).

You should also now be aware of the magnitude of the influence that we, as humans have had – and continue to have – on environments. Despite the tiny proportion of the Earth's history for which humans have been present, our ability to use sources of energy other than the Sun has given us enormous potential to produce change in environments (Chapters Two to Seven). These changes can have devastating impacts on other species and on whole environments and can also have major, uneven effects on humans (Chapter Seven).

○　　Suggest two human actions that are having major environmental effects.

●　　Almost certainly the largest effects are those associated with accelerated climate change, probably due to human use of fossil fuels (Chapter Six). However, human intervention in the hydro-logical cycle in some areas can have widespread effects across national boundaries (Chapter Five). Another major effect over historic time has been the removal of forest cover, as land is converted to agricultural or other uses (Chapter Four).

Even if you were not doing so before you started reading this book, you should by now have begun to question why we do some of these things that have such large effects on environments. A partial answer is to do with economics (Chapters Three and Seven), but there are other factors that are discussed in **Bingham et al. (2003)**, which is the next book in this series. If we are going to act on any of the processes whereby we affect environments, then it is important that we have a clear idea of the extent of change. So the second key question for this book was: how do we determine the consequences and significance of change?

Chapter Two and Chapter Four provide two possible indicators of the significance of change. In Chapter Two the equation $I = P \times R \times T$ is used in order to analyse human impacts; in Chapter Four the ecological footprint is used as a single summary measure linked to the essential resource of land. Both of these examples represent models of the significance of change, so also provide partial answers to the third key question: how can we represent, model and predict changes and their consequences? Modelling and prediction depend on being able to describe situations in an appropriate manner. Through most of the chapters, we have assumed that change was something measurable, and which could therefore be represented by a table of numbers or perhaps a graph. However, the discussion of deforestation in Chapter Four and the debates over climate change in Chapter Six should have alerted you to the fact that measuring change is not necessarily simple nor uncontentious. Measurement takes time and effort, and some of the controversy over environmental change hinges on different interpretations (models) of those measurements that have been made. It is useful to review some of the underlying principles of measurement, to recognize why measurement may be problematical and to draw out more commonalities from the previous chapters.

Questions of measurement

In Chapter Four the decline in the numbers of many familiar bird species in Britain was mentioned. How might we know that such a decline was taking place? People living in rural areas, and birdwatching enthusiasts, would be likely to have noticed what they believed were changes in bird numbers, but is this enough to go on? Observers of the demise of the passenger pigeon (quoted by **Hinchliffe and Belshaw, 2003**) would have been in exactly the same position of uncertainty and it is only with hindsight that the scale of change in numbers of that pigeon species is brutally clear. Faced with a possible ongoing change, how do we measure this, and how might we go about investigating the factors that could be responsible for those changes?

Immediately, we have to be realistic. There is no way that we can count the numbers of all the bird species in which we are interested, over a large area. The most we can do is to obtain a **sample measure** from which we can derive an estimate of those numbers. But before we can even begin to think about how to sample, we have to have some agreed taxonomy of the subjects. Fortunately, the

sample measure

taxonomy of bird species is relatively clear and agreed, but it can be more difficult for other entities or for less well-defined groupings, as we will see in the consideration of systems later in these conclusions.

Birds and other mobile species present particular problems in estimating their numbers. We have to consider questions such as:

- Over what area is it appropriate or realistic to try to estimate numbers?
- At what time in the year is the number to be estimated?
- Is it ethically acceptable to use particular methods of counting the organism?
- What are the sources of error in different methods of counting?

Let's consider these questions for farmland birds in the UK. First of all, how do we define farmland and farmland birds? Since many birds feed in one type of vegetation, nest in another and possibly spend the winter somewhere else entirely, should we relate any estimated numbers to the potential feeding or the nesting area? In theory, it might physically be possible to capture all the birds of the chosen species in a defined area, but how long would this take, and how confident could we be that we had actually captured all of them? Would it be acceptable to kill the birds to be absolutely sure we don't count any one twice, or keep the first caught ones in captivity while we hunted for the rest? In practice, there are accepted techniques for estimating bird numbers, but the numbers *are* only estimates and may be subject to error. Longer-term studies often involve records of sightings by enthusiastic amateurs whom it is hoped will operate in a consistent way.

From this, you should see that any quoted number or change in numbers for 'farmland birds' can only be an approximation. For less familiar species, without the support of enthusiastic amateurs, the situation may be even worse!

Measuring physical attributes of environments is potentially less contentious, but there are still questions of sampling. Think of the rainfall data quoted for Cambridge in Chapter Five. As noted there, rainfall is measured with a rain gauge, and the Meteorological Office in the UK has an agreed standard specification for such gauges. But how representative are the data obtained from rain gauges of the actual rainfall over the Cambridge region? Let's do some simple calculations. Suppose there are perhaps 5 gauges situated within the boundaries of Cambridge. Each gauge has a collecting surface area of a few hundred square millimetres. Even with 5 gauges, the total area over which the rain is sampled is less than one square metre.

○   Assuming Cambridge occupies an area 5 km by 5 km, what is its area in square metres?

●   5 km is 5000 m or 5×10^3 m, so the total area is $5 \times 5 \times 10^6$ m^2, or in standard notation 2.5×10^7 m^2.

If we assume that we can directly measure only the rain that fell over an area of a square metre, this represents only $1/2.5 \times 10^7$ of the total area, or one twenty-five millionth of the total falling over the area of Cambridge. Given that the rain falling on any particular rain gauge will depend to some extent on local air currents, and these will be affected by the presence of buildings or other obstructions, by windspeed and direction, then you may be beginning to wonder how much faith it is possible to place even in such apparently simple measures. In fact, since gauges are likely to be sited so as to minimize such variation, and we have data over relatively long periods, we can be reasonably confident that the data are reliable, but we should always be cautious about accepting numbers at face value, without enquiring how they were determined, and how much faith we can really place in them.

General patterns of change

In addition to the uncertainty that is associated with any data about environments, we also need to consider for any data set its general form of change with time. There are some general patterns of change that we can recognize, and use as a basis for discussion. A frequently quoted example concerns the changes in carbon dioxide concentration measured at Mauna Loa, which illustrates two aspects of this: see Figure 1.

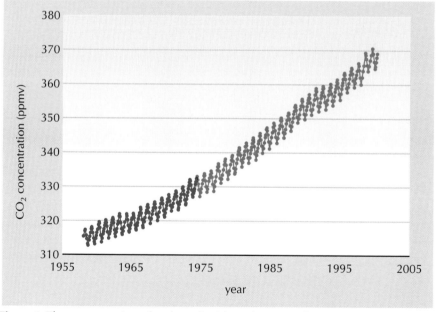

Figure 1 The concentration of carbon dioxide in the atmosphere measured at Mauna Loa, Hawaii.
Note: Data prior to May 1974 (blue) are from the Scripps Institution of Oceanography (SIO); data since May (red) are from the National Oceanic and Atmospheric Administration (NOAA).

Figure 1 actually shows two different forms of change: a gradual increase in the values of carbon dioxide concentration if we consider the whole time-period; and a regular pattern of rise and fall in concentration if we look at the data for just a few years at a time. These two different patterns of change are called:

- monotonic change
- cyclical change.

Monotonic change occurs where a line connecting the data points has a slope that is always either upward, downward or horizontal. Cyclical change is the regular, repeating pattern shown in the detail of Figure 1.

○   Why do the data show the regular rise and fall?

●   The most likely reason is seasonal change, with carbon dioxide being absorbed into the vegetation during the growing season, but being released again or at least not taken up so rapidly during any non-growing period.

○   Identify an example from the description of primary succession given in Chapter One that might show a similar pattern of change to the Mauna Loa data.

●   One possible example is biomass in the early stages of succession in a seasonal environment. This would normally increase overall from year to year, but with a superimposed seasonal pattern, at least until the point where total community respiration equalled primary production.

In truly monotonic change the slope of the graph of whatever is being portrayed either stays the same, or only increases continually or decreases continually. Figure 2 shows this in graphical form.

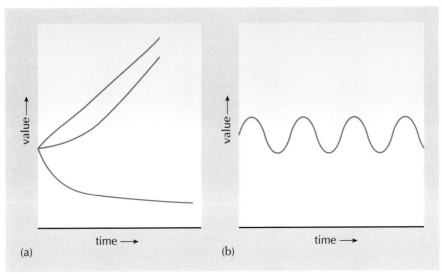

Figure 2 Graphs showing: (a) three possible forms of monotonic change; and (b) cyclical change.

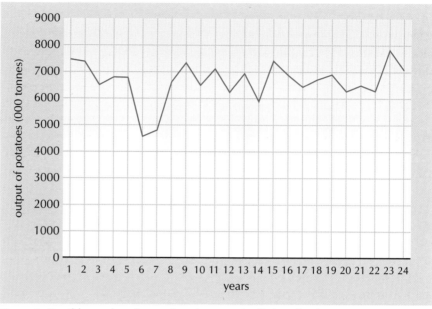

Figure 3 Possibly random fluctuations in some (real) data for the UK potato crop.

To recognize the longer-term monotonic element in the Mauna Loa data, we need to ignore the 'fuzziness' created by the shorter-term variation. Here, it is relatively easy to see this overall pattern, but this is not always the case. Take the example of the changing sea temperature in the North Atlantic, quoted in Chapter Six. This shows a seasonal effect, but any longer-term changes may be much more difficult to detect. Most real environmental data do not show nice clear patterns, and deciding whether something is changing according to a particular general pattern may be quite difficult. Partly, this is due to the superimposition of a third type of change – **random fluctuation**.

random fluctuation

Randomness has a precise mathematical definition but, for our purposes, it is sufficient to regard it as unpredictable change. Where there is a clear monotonic or cyclical pattern, we can make a reasonable prediction where the next point on a graph of the data will appear before we have the actual number. So, if there was a regular pattern like that in Figure 1 we would expect that after a few data points above the general line of points, there would be a few below the line. However, if the pattern were truly random, it might look more like that shown in Figure 3. We cannot say that a high point will always be followed by a lower, higher or similar one.

○   Are there discernible monotonic, cyclic or random aspects to the data shown in Figure 3?

●   I would say there is no clear overall pattern, although if we only had the data in the middle of the period shown, I would be tempted to say there was a regular downward trend coupled with an annual cycle of alternate high and low values. But over the whole period, it looks as if the values just fluctuate in an unpredictable manner about a value of 6.5 million tonnes.

The different possible interpretations of the data in Figure 3 stress the importance of measuring values of any variable of interest over an appropriately long period before claiming that it shows a particular pattern. Different **mathematical models**, like those used for populations in Chapter Three, can be used to generate different patterns of behaviour. These can be explored using computer-based methods, and such studies have led to some interesting new insights about environmental processes, but we should recognize that such models can be misleading. Suppose, for example, that some phenomenon is believed to exhibit monotonic change as a result of measurements taken over a particular timescale. A model could be produced that might predict how this phenomenon will change if the apparently observed pattern continues. However, it is possible that observations over a longer period would show that change was not monotonic, but actually cyclical or random, as in sections of Figure 3. The model predictions, and any actions that were based on these predictions, would then be completely wrong, and could even have the opposite effect from that intended. You may like to think whether the evidence of climate change presented in Chapter Six could be such an example. Certainly, over the period shown in Figure 1, there appears to be a monotonic increase in carbon dioxide concentration superimposed on a seasonal cylical pattern. But over a much longer timescale, that apparent monotonic increase could be part of a long-term cycle. Again, this stresses the importance of time in examining environmental change.

(margin note: mathematical model)

Choosing what systems to model

The second and third key questions are linked, since a major reason for modelling and predicting change is to understand its consequences and significance. However, to be able to model change, we need to decide what it is that we need to model. An idea that has been implicit in much of this book concerns the *linkage* between processes and features of environments. So the movement of materials and energy around ecosystems links together different organisms and parts of the physical environment (Chapter One). The movement of the ocean currents (Chapter Five) affects, and is affected by, the climate (Chapter Six). This interlinkage suggests that we need to consider environments as **systems**, that is, sets of items and processes that are linked together so they influence both one another and the changes in the system as a whole. The word 'system' has become widely used – as in *ecosystems,* the *planning system, hot water systems,* the *legal system* and so on – but it also represents a particular way of viewing the world.

(margin note: system)

There are several discernible traditions within systems thinking: the oldest, based on the work of von Bertalanffy in the early twentieth century, is *systems science*, which uses these ideas to obtain ever greater accuracy and predictive power in models of the way natural systems may work. As an example of this tradition, it has become common to speak of *Earth systems science* (rather than just Earth science) to recognize that the function of the Earth involves linkages between planetary, geological, chemical and biological processes, and it is difficult to understand Earth's history or current behaviour without considering all of these. A related development is *systems engineering*, which uses similar ideas specifically to frame and solve problems, such as the management of some environment.

Both traditions recognize the problem of complexity, in that the more processes that need to be considered, the more difficult it becomes to model them within the available time and knowledge base. To cope with this, we need to consider what is an appropriate **boundary** for the system that we choose to model or manage. Such a boundary defines the limits of the interactions that we consider to be important *for our purpose*. We may be well aware that, for example, change in climate will affect the crops that can be grown in a particular area over a period, but a farmer considering what crop to sow in any particular year is likely to draw a boundary around the farming system of interest that excludes the changing climate system. While such a circumscribing is a valuable practical device, it has its dangers. So, to continue the agricultural example, fertilizer use within a system that only includes the farm and its crops is likely to be considered purely beneficial by the farmer. However, if we draw a wider boundary, to include water courses or even just non-crop species within the farm, then the overall effects of fertilizers may be regarded as less positive.

boundary

○   What effects could fertilizer use have within this wider system boundary?

●   As you remember from Chapters Four and Five, excess plant nutrients in water courses (eutrophication) can increase the growth of undesirable algal species. When these die, the bacteria that decompose the dead algae can remove so much oxygen from the water that fish and other organisms which also need oxygen can begin to suffer.

A similar example was provided by the story of DDT given in Chapter Three. You may like to check back and consider what system boundaries were assumed by the inventors and users of the chemical, compared to the boundary deemed appropriate by those who banned its use.

Both systems science and systems engineering have been greatly strengthened by developments in computing power, and in the theories associated with chaos and complexity. While both these traditions have achieved considerable success, it has been recognized that the purely rational approach which they advocate for scientific and technical problems may be inadequate in dealing with situations involving people, or what can be termed **human activity systems** (Checkland, 1981).

human activity system

The ideas of *interpretive social engineering* developed by Checkland and others emphasize that people's perception and interpretation of situations, problems and data can very strongly affect action to deal with that situation. While emphasizing the need for the techniques and approaches of systems science and engineering, this tradition recognizes that there is rarely one single agreed model of the important system or systems that can be recognized in any situation. Each participant will have a slightly different perception of the situation, and conceptualize it in a different way. So, in our example, the farmer may see the land being farmed as part of a *food production system*, or possibly a farm family *livelihood system*, and this affects what are judged to be desirable changes in the system. The farmer probably seeks to increase production of food per hectare, or economic operating profit, and hence family income. Yet that land is also a component of other systems as perceived by other people. To the casual passer-by, it may be part of a *landscape system*, and a desirable landscape may not be one that is consonant with high levels of production or profitability. To the dedicated birdwatcher, the land needs to be a good habitat for a wide diversity of bird species, and conservationists may be concerned with general biodiversity. However, neither of these aspects is necessarily part of the farmer's view of the land as production system.

Finding ways to act in such a situation may involve trying to draw out these different perceptions, and agree agendas for change that can be accepted by the different parties with an interest in it. Such differences in defining the purpose of environmental systems, and appropriate boundaries for such systems, may underlie some of the contests over environments that are the subject of the next book of this series, **Bingham et al. (2003)**. You may also find it helpful to your understanding of this book to review some of the material in earlier chapters, and to consider what system is defined, or implied, when the author is dealing with particular issues.

References

Bingham, N., Blowers, A.T. and Belshaw, C.D. (eds) (2003) *Contested Environments*, **Chichester, John Wiley & Sons/The Open University (Book 3 in this series).**

Checkland, P.B. (1981) *Systems Thinking, Systems Practice*, Chichester, John Wiley & Sons.

Hinchliffe, S.J. and Belshaw, C.D. (2003) 'Who cares? Values, power and action in environmental contests' in Hinchliffe, S.J., Blowers, A.T. and Freeland, J.R. (eds) *Understanding Environmental Issues*, **Chichester, John Wiley & Sons/The Open University (Book 1 in this series).**

Acknowledgements

Grateful acknowledgement is made to the following sources for permission to reproduce material within this book.

Figures

Chapter One contents page: © Bernard Castelein/Nature Picture library; *Figures 1.1a, b, c, d:* © Aerographica Aerial Photography; *Figure 1.2:* © Hulton Archive; *p.10 (left):* © Raymond Mendez/AA/Oxford Scientific Films; *p.10 (right):* © Alastair Shay/Ecoscene; *p.20:* © Dick Morris; *Figure 1.11a:* © Ecoscene/Sea Spring Photos; *Figure 1.11b:* © Phil Savoie/Nature Picture Library; *Figure 1.14:* © Joel Creed/Ecoscene; *Figure 1.16a:* © Fred Hoogervorst/Panos Pictures; *Figure 1.16b:* © Renee Lynn/Science Photo Library; *Figure 1.16c:* © John Shaw/ Natural History Photographic Agency; *Figure 1.16d:* © Bernard Castelein/ Nature Picture Library; *Figure 1.16e:* © N.A.Callow/Natural History Photographic Library; *Figure 1.22:* Bott, M.H.P. (1982) *The Interior of the Earth, Its Structure, Constitution and Evolution*, Edward Arnold (Publishers) Limited. Reproduced by permission of Arnold; *Figure 1.23:* Fowler, C.M.R. (1990) *The Solid Earth: An Introduction to Global Geophysics*, Cambridge University Press; *Figure 1.24a:* Torsvik, T.H. et al. (1996) 'Continental break up and collision in the Neoproterozoic and Palaeozoic', Earth Sciences Reviews, 40, Elsevier; *Figure 1.24b:* Pharaoh, T.C. (1999) 'Palaeozoic terranes and their lithospheric boundaries within the Trans-European Suture Zone (TESZ)', Tectnophysics, 314, Elsevier; *Figure 1.24c:* Dietz, R.S. and Holden, J.C. (1970) Jour. Geophys. Res, 75, American Geophysical Union; Figure 1.24d: Dietz, R.S. and Holden, J.C. (1970) Jour. Geophys. Res, 75, American Geophysical Union; *Figure 1.25a:* © Peter Scones/Science Photo Library; *Figure 1.25b:* © Ann and Steve Toon/Natural History Photographic Agency.

Chapter Two contents page and Figure 2.12: © Adrian Warren/Ardea; *Figures 2.1, 2.6 and 2.8:* Roberts, N. (1998) *The Holocene: An Environmental History*, Blackwell Publishers Ltd; *Figure 2.2a:* © Peter Steyn/Ardea; *Figure 2.2b:* Bob Gibbons/Ardea; *Figure 2.3a and 2.13:* © Roy Lawrance; *Figure 2.3b: The Patriarch of Grado heals a possessed man* by Vittore Carpaccio, 1494, © Galleria dell' Accademia, Venice/AKG London; *Figure 2.3c:* Courtesy of Barclays Stockbrokers Ltd.; *Figure 2.4:* © Free Library, Philadelphia, PA, USA/Bridgeman Art Library; *Figure 2.5:* © John Downer/Nature Picture Library; *Figure 2.7:* © Vatican Museum Rome/Album/Joseph Martin/The Art Archive; *Figure 2.9:* © The British Library/AKG London; *Figure 2.10:* From *London, a Pilgrimge* by Gustave Doré and Blanchard Jerrold, 1872. © Dover Publications Inc., New York, 1970.; *Figure 2.11a:* © John Hunt; *Figure 2.11b:* © Bruce Davidson/Nature Picture Library; *Figure 2.14:* © Phillips Collection, Washington DC, USA/ Bridgeman Art Library; *Figure 2.19:* Adapted from Bennett, J. W. (1976) *The*

Ecological Transition: Cultural Anthropology and Human Adaptation, Pergamon Press Inc. Copyright © John W. Bennett; *Figure 2.21:* © Martin Bond/Science Photo Library.

Chapter Three contents page and Figure 3.7: © A. Nelson/Dembinsky/FLPA; *Figure 3.1:* Data set collaboration of the Center for International Earth Science Information Network (CIESIN), Columbia University New York; International Food Policy Research Institute (IFPRI) and World Resources Institute (WRI), 2000; *Figure 3.3:* www.multimap.com. © Getmapping plc; *Figure 3.5:* © Ken Wilson/Ecoscene; *Figure 3.6:* © G.Ellis/Minden/FLPA; *Figure 3.7:* © A. Nelson/Dembinsky/FLPA; *Figure 3.10a:* © A.N.T./Natural History Photographic Library; *Figure 3.10b:* © Robert Pickett/Ecoscene; *Figure 3.10c:* © Kelvin Conrad; *Figure 3.13:* © Courtesy of Haileybury and Imperial Service College, Hertfordshire; *Figure 3.14:* Wrigley, E.A. and Schofield, R.S. (1989) *The Population of England 1541–1871: A Reconstruction*, Cambridge University Press; *Figure 3.15:* Wells, H.G. (1925) *The Outline of History*, Cassell & Company Ltd.; *Figure 3.16:* Debenham, F. (1960) *Discovery and Exploration – An Atlas – History of Man's Journeys Into the Unknown*. Copyright © Geographical Projects Limited, London 1960; *Figure 3.17:* © Crown Copyright; *Figure 3.18:* © Fritz Polking/Still Pictures.

Chapter Four contents page: © Still Pictures; *Figure 4.1:* © Still Pictures; *Figure 4.2a:* © Roger Wilmshurst/FLPA; *Figure 4.2b:* © Dunbar Arnold/Agstock/SPL; *Figure 4.2c:* © L. Lee Rue/FLPA; *Figures 4.3a and b:* © Jeremy Horner/Panos; *Figure 4.3c:* © Ed Young/Agstock/SPL; *Figure 4.4 (left):* © Ardea; *Figure 4.4 (right):* © Paul Smith/Panos; *Figure 4.6:* © Tantyo Bangun/Still Pictures; *Figure 4.7:* © USGS Survey, courtesy of Infoterra Ltd.; *Figure 4.9 (both):* © Bob Gibbons/Ardea; *Figure 4.10 (both):* © Mike Dodd; *Figure 4.11 (left):* © Popperfoto; *Figure 4.11 (right):* © Jim Holmes/Panos; *Figure 4.14a:* © Mike J. Thomas/FLPA; *Figure 4.14b:* © E & D Hosking/FLPA; *Figure 4.14c:* © Bob Gibbons/Ardea; *Figure 4.15a, b, c and d:* © Stephen Trudgill. From: Field Studies Council, 'Soil types', *Field Studies*, vol.7, no.2 1989; *Figure 4.17 (left):* © Courtesy of Philip Owens; *Figure 4.17 (right):* © R.J.Godwin/Cranfield University, Silsoe with kind permission; *Figure 4.18:* Jenkinson, D.S. (1990) 'The turnover of organic carbon and nitrogen in soil', *Philosophical Transactions: Biological Sciences*, vol.329, p.365, The Royal Society; *Figure 4.20:* © Jim Gipe/Agstock/SPL; *Figure 4.21:* © Courtesy of Tim Hess; *Figure 4.22:* © Department of Agriculture, Victoria. From: Neil Barr and John Cary, *Greening a Brown Land*, Macmillan, 1992; *Figure 4.24:* © Science Museum/Science and Society Picture Library; *Figure 4.25a:* © Tony Craddock/SPL; *Figure 4.25b:* © Maurice Nimmo/FLPA; *Figure 4.25c:* © Courtesy of Kennecott Utah Copper Corporation; *Figure 4.26:* © Denis Wirrmann/I.R.D. Photothèque Indigo; *Figure 4.27:* © Chris Mattison/FLPA; *Figures 4.28 and 4.29:* Wackernagel, M. and Rees, W. (1996) *Our Ecological Footprint*, New Society Publishers.

Chapter Five contents page: NASA Human Space Flight Gallery; *Figure 5.1:* NASA Human Space Flight Gallery; *Figure 5.2:* © Charles Darwin Foundation, Galapagos; *Figure 5.4:* © EPA/PA Photos; *Figure 5.6:* Adapted from www.met-office.gov.uk/climate/uk/averages/images/rainannuals6190.gif. © Crown copyright, Met Office. Reproduced under licence number MetO/IPR/2; *Figure 5.11:* IPR/32-7 British Geological Survey © NERC. All rights reserved; *Figure 5.16:* © T.C Middleton/Oxford Scientific Films; *Figure 5.17:* Chorley, Schumn and Sugden (1985) Geomorphology, Methuen & Co.; *Figure 5.18:* © Popperfoto; *Figure 5.20:* Reproduced courtesy of the Library and Information Centre, Royal Society of Chemistry; *Figure 5.21:* Adapted from Hulme, M. and Jenkins, G. (1998) UKCIP's Technical Report Number 1, September 1998; *Figure 5.22:* © John Giles/PA Photos; *Figure 5.23:* © David Sandwell, Scripps Institution of Oceanography; *Figure 5.24:* Illustration by Robert Dudley from W.H.Russell, 'The Atlantic Telegraph', Day & Son Ltd., London, 1865; *Figures 5.25a and b:* © Pat Brandow, NOAA; *Figure 5.26:* © W James Ingraham, NOAA/Alaska Fisheries Science Center; *Figure 5.27:* © David Walker Photography; *Figure 5.31:* Image provided by ORBIMAGE. © Orbital Imaging Corporation and processing by NASA Goddard Space Flight Center; *Figure 5.33:* © B. Murton/Southampton Oceanograpy Centre/Science Photo Library; *Figure 5.34:* Cartoon by Illingworth, *Daily Mail*, 1st January 1953. Copyright © Associated Newspapers/Atlantic Syndication; *Figure 5.35:* Adapted from www.fao.org/NEWS/FACTFILE/ff9604-e.htm, Food and Agriculture Organization of the United Nations; *Figures 5.36a and 5.40:* © Mark Brandon; *Figure 5.36b:* © M. Watson/Ardea; *Figure 5.37:* © Martin Bond/Science Photo Library; *Figure 5.38:* © Science Photo Library; *Figure 5.39:* © Shetland Islands Council.

Chapter Six contents page: © NASA/JSC Digital Image Collection; *Figure 6.1:* © NASA/JSC Digital Image Collection; *Figure 6.2:* © Mike E. Dodd; *Figures 6.8, 13, 14, 18, 20, 21, 23, 25:* Adapted from Climate Change 2001-1, Intergovernmental Panel on Climate Change; *Figure 6.9:* © Adam Hart-Davis/Science Photo Library; *Figure 6.10:* Introduction to Climate Change, UNEP/GRID, Arendal, Norway; *Figure 6.11:* © C.Clifton/Minden Pictures/FLPA; *Figure 6.12:* © British Antarctic Survey/Michelle Gray; *Figure 6.22a:* © Pierre Laboute/I.R.D. Photothèque Indigo; *Figure 6.22b:* © Georgette Douwma/Science Photo Library; *Figure 6.24:* © EPA/PA Photos.

Chapter Seven contents page and Figure 7.3: © Dr Morley Read/Science Photo Library; *Figure 7.1:* Radford, T. (2000) 'Greenhouse build-up worst for 20m years', *The Guardian*, 17 August 2000. Copyright © The Guardian; *Figure 7.6a:* Courtesy of British Antique Replicas Ltd.; *Figure 7.6b:* © Mark Edwards/Still Pictures; *Figure 7.8:* Courtesy of Stopesso; *Figure 7.9:* © Jean Pierre Dutilleux/Reuters/Popperfoto; *Figure 7.11a:* © National Archives of Canada/PA-051492; *Figure 7.11b:* © National Archives of Canada/PA-051498; *Figure 7.13:* OU Library Archive; *Figure 7.14:* © Eamonn McCabe/The Guardian; *Figures 7.16a and b:* © Jorgen Schytte/Still Pictures.

Tables

Table 1.1: Terborgh, J. et al. (2001) 'Ecological meltdown in predator-free forest fragments'. Reprinted with permission from *Science*, vol.294. Copyright 2001 American Association for the Advancement of Science; *Table 3.1:* UN (2000) *Population Reports*, vol. XXVIII, no.3, Series M, no.15 Special Topics, United Nations; *Table 4.5:* Pretty, J.N. and Howes, R. (1993) 'Sustainable Agriculture in Britain: Recent Achievements and New Policy Challenges' (*Research Series* vol.2 no.1), International Institute for Environment and Development; *Table 6.4:* Adapted from Climate Change 2001-1, Intergovernmental Panel on Climate Change.

Cover illustrations

From left to right: © Dick Morris; © Dick Morris; © Photodisc; © Mark Brandon.

Every effort has been made to trace all copyright owners, but if any have been inadvertently overlooked, the publishers will be pleased to make the necessary arrangements at the first opportunity.

Index